U0169500

食 物 小 传

派

Pie

A Global History

〔澳大利亚〕珍妮特·克拉克森 著

李天蛟 译

北京联合出版公司
Beijing United Publishing Co.,Ltd.

目 录

对派的初步观察

> 我可能无法对派进行准确定义，但当我看到的
> 时候会知道那个东西就是派。
>
> ——雷蒙德·索科洛夫

在开始创作这本书之前，我曾被告知，每一位作者都
应该在正式动手写作之前充分考虑作品内容所包含的范围。
当时我觉得，没有什么问题。不就是关于派嘛，很简单。
另外我还被提醒道：作者必须在作品开头描述一下这部作品
的内容范围，以便预先告知读者。当时我在想："这个要求
很简单，我在作品开头为派下个定义就可以了。"结果等我
真正开始着手的时候却一败涂地。如果我能早一点想起本
章节开头的那句引语，大概就不会为这次失败感到如此痛
苦。如此博学出众的雷蒙德·索科洛夫尚且无法对派进行
定义，我又有什么资格狂妄地进行尝试呢？不过我并没有
为这次尝试感到后悔，因为它出乎我意料地吸引人，给我
带来了全新的启发。

关于派，我首先想到的是曾经在一家高档餐厅点的一份鹿肉派。一段预料中的等候时间过后，我的派被端了上来。这份派看起来十分精美，松软的蒜香土豆泥做底，蔬菜丝组成巧妙的分层结构，顶上是鲜嫩多汁的半生鹿肉片，整体由一个半圆拱形的千层酥皮包裹着。它十分美味，但是我非常确定，这并不是真的派。对于派的问题，我和这家餐厅的厨师之间显然存在很大的意见分歧。不过，这种分歧也使我顿悟：即使我对派做出了简洁准确的定义，世界上肯定仍然有人无法同意我的观点。

但一定可以找到一些存在共识的角度作为出发点吧？派之所以会成为派，肯定不仅仅是因为人们把它称作"派"。例如，美国有一种"爱斯基摩派"，这个东西实际上是一种冰激凌；"月亮派"则是一种打着派的幌子的巧克力饼干；"波士顿奶油派"其实是一种蛋糕——很明显，这种蛋糕之所以被称为"派"，是因为烘焙过程中使用了制作派的烤盘。按照这种逻辑，我用制作派的平底锅熬一锅粥，是不是就能叫作"粥派"了呢？其他类型的"派"更是成问题。香肠卷算不算一种小型的派呢？我可以把厚顶水果挞①和苹果饼②当作派吗？它们看起来像是油酥面皮破掉的失败的派。传统的苏格兰黑面包和英国的重油蛋糕都是有油酥面皮外壳的水果蛋糕，那么这两种食物可以算作派吗？我还想到

---

① 原文为 cobbler，一种用水果制成的甜点，顶部覆盖一层较厚的饼干面团。——编注
② 原文为 pandowdy，一种甜点，将黄油、牛奶和鸡蛋的混合液灌入装有苹果和香料的烤盘烤制而成。——编注

了农舍派和牧羊人派，这两种"派"都有与派类似的分层结构，然而（对我来说）并不算真正的派。为什么不是呢？第二次顿悟告诉我，因为这两种食物都没有油酥面皮。这样一来，至少就本书的写作目的而言，我对派进行定义的尝试似乎演变成了一套最低标准。第二次顿悟帮助我总结出了"派的第一定律"："无油酥面皮，不成派"。

出于第一章内容将提到的一些原因，我又马上想到了"派的第二定律"：派必须是烤制而成的，通过油炸、蒸煮等方式做出来的食物都不是派。接下来，只要再找到一条关于外皮数量和位置（单层顶、单层底或者双层皮[1]）的定律，我的这套标准就可以成形了，至少我当时是这样认为的。然而制定"派的第三定律"确实特别困难，而且我至今仍无法确定自己是否采用了正确的方式来制定这些标准。于是我开始到《牛津英语词典》中寻找有关外皮的信息。《牛津英语词典》的内容显示，派是：

> 由水果、肉、鱼或者蔬菜烤制而成的食物，顶面覆盖着油酥面皮（或者其他类似食材），底部和侧面常常也有油酥面皮；也指（主要在北美）填满水果烤制而成的开放形态的油酥面皮食物，即水果挞或水果馅饼[2]。

---

[1]　即顶面和底面都有派皮包裹。——编注
[2]　原文为 flan，一种底部和侧边包有油酥面皮或海绵蛋糕的点心，馅料可以是水果或咸口食材。——编注

这样看来，《牛津英语词典》的诸位编辑认为，真正的派必须要有顶面油酥面皮，底面油酥面皮可有可无，而美国是个例外，在美国最重要的是底面油酥面皮。词典中的内容似乎在暗示，带有底面油酥面皮的美国派，按照英式英语，相当于水果挞或者水果馅饼。如果他们进一步把水果馅饼解释成"开放形态的水果挞"，把水果挞解释成"与派相同或相似的一种食物"，那么这里的定义就会更加含糊不清。

我对《牛津英语词典》一向怀有热爱与崇敬之情，但与此同时不得不说，这部词典与大多数说英语的国家和地区类似，对派和水果挞之类的东西缺乏清晰的界定。1927年，《泰晤士报》的通信版面上曾爆发了一场非同寻常的争论，有力地印证了这一点。1927年9月，一位名叫R. A. 沃克的先生激动地给《泰晤士报》写了一封信，对一家餐厅表达了"强烈的抗议"。他表示，这家餐厅"打着苹果挞的旗号，提供的却是千层酥皮和炖水果，这种骗人的把戏令人发指"，为此，他强烈谴责这家餐馆的老板。沃克先生在这里指出的是苹果派的质量问题，但很快因为"把苹果派和苹果挞混为一谈"而遭到了约翰·萨默维尔上校的指责。上校表示"一刻都无法容忍"这种混淆现象，还声称"任何受到良好教育的孩子"都可以区分这两种东西。上校认为，"派"是一种"表面覆盖着油酥面皮的食物，油酥面皮下面可以使用任何食材""只有水果暴露在底面油酥面皮

4

上的时候才能叫作水果挞"。于是，接下来的两周时间里爆发了惊人的激烈争论，人们进行了隐晦的侮辱和明确的陈述，还抛出了一些其他次要问题。一些移居海外的美国读者和法国读者随后也加入了讨论。随着争论进入白热化阶段，亨利·爱德华·阿姆斯特朗教授作为一名"简明命名法的拥护者"现身说法。他表示，自己"由衷支持"这一类讨论，接着还举了一个他眼中最好的例子："米派的重点永远在于'派'。"据阿姆斯特朗教授的描述，"把米连同适量白糖和足量的农家牛奶放进烤盘里"进行烘烤，就可以制成派，"用科学界的行话来讲，派的表面会精妙地自动结出外皮"。他还澄清这种食物不是布丁，因为"布丁必须包在布料里进行水煮，也有外皮，但外皮不会出现棕褐色痕迹"。阿姆斯特朗教授没有提及自己的专业学术领域，不过他那略带自负的观点引发了一位牛顿先生的愤怒回应。这位牛顿先生对阿姆斯特朗教授的专业身份提出了质疑，声称阿姆斯特朗教授"肯定不是一位烹饪教授"。牛顿先生指出，教授提到的"外皮"只不过是烤箱加热时产生的一层表皮，而派的"外皮"是由"与馅料完全不同的食材制成的"。另外一位受到教授观点冒犯的来信者质问阿姆斯特朗教授这位"老顽固"，如何对肯特郡那道著名的"布丁派"进行分类——"布丁派的底部和高高的侧边都有油酥面皮……制作的时候需要把蛋奶酱添加到油酥面皮边缘的高度，随后放入烤箱进行烘烤"。这位来信者同时告诫道："如果你想在肯特郡吃到这道菜，就不要点蛋挞。"

亚伯拉罕·博斯的作品《糕点店铺》。注意图中挂在钩子上的派模具。

在由衷支持与愤怒反驳的交锋中，《泰晤士报》的职员们为"那些认为派和水果挞不是同一种东西的人"发表了一篇简短的社论。这篇社论承认"伟大的牛津词典"和福勒先生关于现代英语用法的书都"放弃"了区别派和水果挞这两种概念，同时温和地批评了阿姆斯特朗教授，认为他以一种无益的方式把布丁拉进了这场争论。位于斯特兰德街的辛普森餐厅是英国美食的代表性餐厅，因此这场争论征询了该餐厅经理的观点作为"权威支持"。这家餐厅的经理赫克先生表现出了出人意料的谦逊姿态，他表示自己并不想在信件中说太多，只是"提出建议，或更准确地说是恳求——英国人应该尽全力抵制当今为传统英文词汇赋予全新含义这一趋势"。他提及了查尔斯·狄更斯,说这位绅士若还在世,定会对他所说的派毫无质疑,即派是一种"将馅料与表面金黄色油酥面皮一同在烤盘内烹制而成的美食"。

一周之后,休·温德姆夫人引用了疑似格拉德斯通[1]的权威观点。温德姆夫人声称自己曾听到他"热情又优雅地对这个话题发表了观点",而他的观点显然是:"派的馅料使用的是水果,使用到肉类的应该称之为馅饼。"这种观点本来有可能开启关于派和馅饼的新一轮讨论,然而《泰晤士报》的读者或许已经厌倦了有关话题的争论,上述观点并未引

19世纪晚期制作西梅干挞的场景

---

① 此处的"格拉德斯通"指的是威廉·尤尔特·格拉德斯通(William Ewart Gladstone, 1809—1898),英国政治家,自由党人,曾四次出任英国首相。——译注

起反响。《泰晤士报》收到的最后一封信来自一位弗兰克·伯德伍德先生，他在信中笃定地声称，英文中的 tart（水果挞）和 torture（扭曲、折磨）这两个单词来自相同的拉丁语词根，"'挞'自然得名于这种食物扭曲的油酥面皮，油酥面皮一般用来装饰水果挞使用到的果酱或其他主要原料，整个水果挞具有一种'开放'的形态。"

同年 10 月 12 日，一篇社论宣布英国进入官方的"布丁季"，正式结束了这场"乐趣满满的水果挞与派之争"。看到英国民众仍然在大量食用布丁（有甜有咸），《泰晤士报》的编辑们似乎松了一口气。因为这种现象显然表明："无论如何，在大英帝国的中心，并没有完全丢失我们古老的坚毅品格。"

那位在《泰晤士报》的争论中赞扬米派的阿姆斯特朗教授其实可以在 1910 年美国最高法院做出的一项判决中找到论据来证明米布丁是一种油酥糕点。当事人约翰·米诺罗普洛斯把一家店铺的门口一部分租给了另一位当事人约翰·科瓦托斯，根据租赁协议，科瓦托斯不得出售特定的几种物品，其中包括"任意种类的油酥糕点"。然而一段时间之后，科瓦托斯开始出售西梅干和米布丁，并在接下来的 47 天时间里售出了将近 3000 盘布丁。因此，米诺罗普洛斯提起诉讼。一审法官认为，西梅"有着低调却美好的历史，默默无闻，值得尊敬"，于是立即驳回了关于西梅干的索赔，把有关米布丁的问题交给了陪审团。陪审团认定米布丁属于油酥糕点，于是科瓦托斯提起上诉。尽管科瓦托斯的辩护律师发表了"言辞激烈的演说"，引用了很多字典里的有关内容，并针对米布丁性质的共识问题展开了讨论，但最终仍然败诉。法院判处科瓦托斯向米诺罗普洛斯支付101.05美元的损害赔偿和21.5美元的诉讼费用。

如果《牛津英语词典》、19世纪博学的《泰晤士报》读者和美国最高法院尚且无法针对"派到底是什么"这个问题达成共识，那么就本书而言，我完全可以对派的外皮自由定义。从历史上看，派的重要特征之一在于可以用手拿着吃，所以大致可以排除那种必须用深盘盛放的"锅派"（pot-pie）。曾有一段时间我常常使用这种限制条件，但后来又因常识和日常惯例对此进行了否定。提出的"第三定律"似乎有一点多余："派可以使用任意数量的外皮，但外皮必须（如第一定律所述）是油酥面皮。"这里基于一些真实可信的历史原因强调了油酥面皮，而不是"其他类似食物"，具体原因将在第一章的内容中阐明。

单词往往可以提供有关事物起源的线索，我希望多了解一下派的历史，于是就去查了pie（派）这个词。《牛津英语词典》显示，这个词首次出现在1303年约克郡博尔顿修道院的会计账目里[虽然读音相近的姓氏Pyman（派曼）最早出现于1301年]。不过与此同时，牛津词典承认这个词的起源不明，并且"英语之外其他语言中的相关词汇未知"。根据牛津词典，pie这个词还可以表示喜鹊（magpie），而对于magpie这个词，"很多观点认为其词源与pie存在关联"。这里"存在关联"的意思是，派的馅料具有"混杂"

的特征，与喜鹊的毛色以及喜鹊为了装饰鸟巢而收集的零碎物品有相通之处。为了支持这一论点，《牛津英语词典》还指出了 haggsi（苏格兰肉馅羊肚，有混杂的馅料）与古法语单词 agace、agasse（喜鹊）之间的相似点。

《牛津英语词典》提出了另一个具有可能性的起源，即 pie 这个词代表一种由油酥面皮烘烤制成的食物，但是也可以用在农业情景之下，表示"堆成一堆的东西"，比如土豆或其他使用土和稻草埋起来储存的农产品。除此之外，这部词典可能还暗含了另外一条线索：英语中的方言单词 piggin 可表示木桶或者用来做饭的陶土锅，这个词可能与盖尔语中的 pige（或其变体之一）存在关联。我并不是语言

维多利亚时代的一幅版画，画中的孩子在用很大的力气切派。

学方面的专家，但是如果我可以发表看法，这里是否也与 pie 之间存在关联呢？本书第一章将针对派的早期历史展开探讨，届时我们可以发现，上述关联的可能性并不亚于派与喜鹊之间存在的关联。

我们同样无法排除这个问题与法语的联系。1066 年诺曼人入侵英国之后，英语发生了翻天覆地的变化。英语和法语中存在很多相似的词汇，这说明一些词汇的起源相近，比如英语中的 tart（水果挞）和法语中的 tourte（圆馅饼）。我们在后续章节中也会遇到其他类似的单词，其中最有趣的是 paste（面糊）、pastry（油酥面皮或油酥糕点）和 pasta（意大利面），这几个单词都是"面粉和水"这一类词义的变体，有着相同的词源。法语中的 pâté（派）也有相同的词根；这个单词中字母 a 上方的符号为长音符，长音代替了 pastry 中的字母 s，这说明法国的派曾经也有油酥面皮。现在的法国派没有油酥面皮，所以最原始的那种带有油酥面皮的派现在被称为油酥面皮派(pâté en croûte)。最有趣的是，pasty（肉馅饼）也有相同的词根，那么问题来了：为什么肉馅饼这种完全由油酥面皮包裹的食物，其名字的起源与普通的"派"的起源毫不相干？

派的简史

在时间和各类开支的作用下，派不再是一块块不易消化的面团。

——《一首关于苹果派的诗》（威尔斯特德，1750 年）

很久很久以前，人们使用烤炉烤制的所有食物，除了面包都叫作"派"。具体原因且听我慢慢道来。

最早的烤炉实际上是一种窑炉，一般用来烧制雕像、陶罐等黏土制品。当时的面包是在炉石上烤制而成的，形状扁平。后来，也许在某个地方，有人发现黏土和面团特别相似，或许出于一时兴起，也可能通过了周密的实验，便把面包放进窑里面烘烤，由此便诞生了最早的面包烤炉。在当时，人们会将肉用扦子穿住放在火上直接烤，或者直接把肉放在生火的煤堆里烤熟。这种烤肉方法的问题在于，即使没有烤焦，也会流失宝贵又美味的肉汁，导致肉块缩水变干。为了解决这个问题，一些厨师开始用树叶或者黏土等材料把肉包裹起来进行烤制。而在另一时间另一地点，可能又有位厨师觉得黏土与面团相似，于是就出现了最早的肉派。中世纪的厨师称之为"烤物"（bake-mete）[1]。

早期，派表面的厚皮发挥了烤盘的作用。后来的几百年时间里，人们烤制食物的唯一容器就是这种外皮，因此所有这些食物都属于派。事实证明，外皮还具有另外两个有用的功能：它发挥了外包装的作用，便于携带和储存（当

---

[1] mete 在中古英语中意为"食物、营养"。——编注

一家法国面包店。图片出自《牧羊人的日历》（约 1499 年）。

时还没有出现午餐盒）；另外，包裹外皮还可以排除空气，因此在罐装与冷藏技术出现之前，外皮还起到了保鲜的作用。当时的人们把早期的派皮叫作 coffin①，这在今天听起来有些阴森，先辈似乎已在通过这种名称暗示派的用料值得怀疑。实际上，coffin 这个词最初的意思是篮子或者盒子（想想一个香水套盒②），在成为丧葬用语之前可以用来指点心盒。

那么这种外皮吃起来怎么样呢？当时烤制食物的过程

<hr>

① coffin 在现代英语中意为"棺材"，但在中古英语中意为"盒子、篮子、派皮"。——编注

② "香水套盒"的原文为 coffret of perfume，其中 coffret 的意思便是"套盒"。——编注

需要耗费几个小时的时间，所以这种外皮一般都有几英寸厚，而且很硬，这样才能承受长时间的烘烤。现代文献通常认为这种外皮不可食用，而且不是用来吃的。这种含糊的说辞在我看来缺乏可信度。对于我们这些过上了精致生活的现代人来说，这种外皮听起来确实不会好吃，然而在那个艰难的时代，种植、收割谷物以及制作面粉（即便是非常粗糙的面粉）都需要付出难以想象的大量劳动。面粉在当时得来不易，把包裹的馅料吃掉之后，想必人们不太可能会扔掉面粉制成的外皮。即使在权贵家里也是如此吗？当时的勋爵和贵妇可能不会吃这种外皮，但是他们需要养

一场 15 世纪的法国宴会的情景。图片出自让·富凯的《关于兰斯洛特的叙事诗》。

活自己的仆人，人们也指望着他们施舍当地的穷人，因此他们可能会把这种浸透了肉汁的外皮分给厨房里帮厨的小孩和门外乞食的人。

早期手稿中有迹象表明，当时的人们其实会对这些外皮进行重复利用，至少偶尔会这么做。15 世纪一份有关"七鳃鳗派"的手稿显示，吃完七鳃鳗以后，可以将剩余的汤汁连同葡萄酒、糖以及香料一起煮沸，然后倒进铺了几层白面包的外皮里，这样就制成了一道叫作"泡肉冻"的新菜。有时候人们也会把这种烤过的外皮回收，作为增稠剂使用。比如 17 世纪的一份"西班牙杂烩"菜谱提到，添加"一些面包皮或者鹿肉派皮"，然后把炖菜"以小火慢炖五六个小时"。也就是说，当时的人们曾经把这种外皮当作如今的油面糊①使用。

幸好，接下来我们会看到，怎样处理剩余的坚硬外皮这一问题最终得到了解决。

油酥面皮的发明

经过数千年的探索，原始的黏土烤炉终于变成了今天那闪亮的钢制烤箱。然而，只有高科技烤箱还不足以把"烤肉"变成我们所熟知的派。这里还需要另一项更重要的发明——油酥面皮。

---

① 油和面粉加热混合而成的增稠剂，多用于法餐。——编注

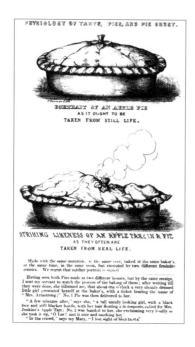

苹果派。图片出自亚力克西斯·索耶的《现代家庭主妇》。

　　给面团加入油脂之后就可以制成油酥面皮。但油酥面皮对于面团、油脂以及二者的混合方式都有特殊要求。在留存至今的最早的食谱书出现之前，人类就已经制作出了油酥面皮，所以我们永远无法得知发明油酥面皮的确切时间、地点和方式。但我们仍然可以做出一些有根据的猜测，其中一些线索来自关于面团与油酥面皮之间关键性区别的思考。为此，我们需要简单介绍一下食品化学领域的知识。

　　这里的关键是一种叫作"麸质"的物质。麸质是一种蛋白质，其分子较长且具有弹性，可以通过提供特殊的结构使面团更结实，也可以锁住气孔中的空气使面团更轻。大量的麸质可以使面团结构牢固，这种情况有利于做面包，

拉瓦雷纳所著《糕点师弗朗索瓦》（1655 年）的扉页

但不适合做油酥面皮。相反，麸质太少的话就无法形成牢固的结构，也无法锁住空气，做出来的面包很扁，油酥面皮很硬。制作油酥面皮的厨师需要把麸质控制在最佳含量，这样做出来的油酥面皮才能既轻薄又酥脆。那么他们是如何做到控制麸质含量的呢？

小麦是唯一一种含有大量麸质的谷物，这样一来我们就得到了油酥面皮起源的第一条线索：上等的油酥面皮只能发展于小麦产地。黑麦、大麦和燕麦无法做出质量较好的油酥面皮，大米、玉米和土豆淀粉同样不适合做油酥面皮。厨师可以通过多种方法控制小麦粉面团的最终麸质含量。具体的麸质含量取决于面团用途，结实的面包与轻薄的酥皮所需含量各不相同。厨师制作油酥面皮的技巧又为我们提供了有关油酥面皮自身发展轨迹的其他线索。

技巧的第一点在于恰当的用水量，水分可以激活面粉中的麸质。第二点在于油脂，油脂可以从多个方面改善油酥面皮的质地。油脂把一小团一小团的面粉包裹起来，可以起到防水的作用，限制进入的水分含量。水分越少，麸质就越少。油脂还可以使麸质链变得"短"且光滑。另外，小片或小团的油脂能够把一小层一小层的面团隔开，使其形成单独的薄片和碎屑，而不是一块实心面团。由此可见，油脂与面粉的比例是决定油酥面皮最终质地的关键因素。但这并非唯一的因素。

油脂与面粉的道理相同，并非所有的油脂都可以产生相同的效果。植物油在室温环境下为液态，看起来似乎与动物油没有什么其他本质差别，但使用植物油无法做出质量较好的油酥面皮，因为在制作过程中面粉会吸收这些油，形成粉质状而不是又软又薄的面团。最适合制作油酥面皮的是熔点较高的脂肪。脂肪融化所需时间越长，小块面团之间的隔离时间就越长，这样有助于面团内部产生蒸气，使面团膨胀、变轻。猪油熔点高，含水量极低，这两点都非常适合制作油酥面皮；而黄油熔点低，在人体体温环境下就开始融化，无法帮助形成上等的质地，但使用黄油可以使油酥面皮的口味更加浓郁。那么在这里，我们就得到了关于油酥面皮发源地的第二条线索：那些真正可口、轻薄且松脆的油酥面皮只能发展于固态脂肪充足的地区，也就是饲养猪或者牛的地区。

　　制作油酥面皮是让每一个业余面点师感到头疼的事情。因为除较多的原料细节之外，油酥面皮同样非常考验技术。众所周知，制作油酥面皮通常需要保持双手、用具以及厨房环境温度较低，用水量越少越好，操作的时候还要轻拿轻放。了解了这个过程之后，其背后的基本原理是显而易见的。在凉爽的环境下操作可以延长脂肪在面团中维持固态的时间；最少的用水量可以减少麸质含量，提升酥脆口感；操作的过程中将力度控制在最小，同样可以降低麸质含量，所以制作油酥面皮时揉面团的方法与制作面包时不同。

　　根据上述发展的时间框架，我们大概可以合理推断，派出现在 14 世纪之前种植小麦并且饲养了猪和牛的某些欧

15 世纪的意大利壁画，画中的烘焙师正在使用长柄铲子把派放入烤炉。

16世纪80年代的布面油画《厨房》，文森佐·坎皮绘。

洲地区。毫无疑问，在整个中世纪时期，北欧和中欧地区有着种类繁多的派。出现这种现象的另一部分原因在于，南欧地区森林资源耗尽之后的很长一段时间，北欧的森林资源仍然可以提供充足的燃料以及饲料。为什么这点很重要呢？因为烤制大体积的"烤肉"与派（里面可能包含整块的鹿腰腿肉）需要花费大量时间和大量燃料。

　　当然，尺寸并不是派的唯一重要因素。油酥面皮艺术的另一分支衍生出了姿态优雅的水果挞以及其他小型油酥糕点，特别是那些甜口的点心，令大多数人无法抗拒。小型甜味油酥糕点的基本概念似乎是在7世纪阿拉伯帝国扩张时期自阿拉伯世界传到了欧洲；不过很可能在文艺复兴时期，意大利北部地区的人们对其进行了改良与发展，制成了我们如今喜爱的众多美妙糕点。意大利北部地区拥有各

15世纪晚期《牧羊人的日历和堆肥》中的木刻版画。中世纪的宴席经常会出现裸浴。画中有一份立式派①，供这对夫妇沐浴嬉戏之后享用。

类油酥面皮糕点发展所需的资源与环境。这里的小麦比较"软"，也就是说麸质含量较低，适合制作油酥面皮。黄油是意大利北部地区最常用的烹饪油脂，同时在当地也是财富的象征；与意大利南部地区使用的油相比，毫无疑问，黄

---

① 原文为 raised pie，指高度较高且无须额外支撑便可独自立住的派，此类派的派皮多为烫面酥皮。——编注

油更适合用来制作甜味派和水果挞中质量上乘的油酥面皮。文艺复兴时期的哲学环境鼓励了各种艺术的发展，意大利王朝家族在享受生活方面也向来豪掷千金，在这种背景下，那些具有创新精神的优秀厨师受到了鼓励与青睐，烹饪领域的理念和技术也开始蓬勃发展，逐渐传播到了法国及欧洲其他地区。

英国的情况与意大利不同。在当时的英国，黄油属于穷人的食材，富人更喜欢猪油，原因可能在于获取猪油需要屠宰牲畜，因此猪油被认为具有更高的感知价值。猪油可以用来制作那种填满肉馅的巨大的立式派，这种肉派得到了蓬勃的发展，成了英国烹饪界的瑰宝之一。

中世纪的油酥面皮是否可食用是一个具有相对性的问题。在这里我们真正关心的是，当时有多少油酥面皮是可食用并且值得食用的。最早的一批食谱中就有暗示，至少在某些情况下，油酥面皮是用来吃的。否则食谱作者为什么强调油酥面皮要做得"尽可能薄且柔软"呢？1545年，英国伦敦出版了一本小书，名为《一本专业的烹饪新书》。上述指导意见就出自该书中的一个食谱，而这也是人们已知的最早书面记载的油酥面皮食谱：

如何制作水果挞的简易油酥面皮[1]

在精细面粉中加入适量的水、一些甜黄油、少

---

[1] 原文为 short paest for tarte，是油酥面皮中最基础、制作方式最为简单的一种，也是最常用的派皮、挞皮。——编注

量藏红花以及两个蛋黄，把油酥面皮做得尽可能薄且柔软。

　　对于食物历史学家来说，早期食谱中的指导内容缺乏细节，非常令人头疼。这些指导往往建立在假定的知识基础之上，更像是供那些有经验的厨师使用的备忘录。不过很明显，到了 16 世纪，简易油酥面皮和千层酥皮，甚至可能还有泡芙酥皮的制作方法都已定型，而且还有迹象表明，截至 16 世纪，这些油酥面皮可能已经存在了几百年时间。英国最早的食谱书名为《烹饪方法》，由英格兰国王理查二世的主厨们于 1390 年左右编写而成。这本书中提到了"千层包"的制作说明，而"千层包"听起来很像是千层酥皮。

一份梅尔顿莫布雷派

有一种特殊的油酥面皮在派的历史中扮演了重要角色，打破了制作油酥面皮的一些方法规则。这种油酥面皮跨越了面包面团与常规酥皮面团之间的桥梁，今天我们仍然可以在一些英国传统猪肉派里看到它。比如莱斯特郡梅尔顿莫布雷镇著名的猪肉派，它的油酥面皮就是我们今天所谓的"烫面酥皮"。烫面酥皮面团与面包面团以及其他油酥面皮面团相比，最大的优势在于可以像黏土那样进行塑形，做成那种可以独立支撑的高外皮。高外皮的出现带来了更多的可能性，因为这种外皮的内部可以填充炖菜、水果以及蛋奶酱等液态形式或者汤汁较多的馅料。

在烤盘出现之前，人们把各种形状的面团作为容器来烤制食物。所以从定义上来讲，当时使用烤炉烤制的各种食物都可以叫作"派"。这种烹饪传统在一部分食物的名称上得到了延续，而且这些食物名称乍一看与派没有任何关联。英文中的 custard（蛋奶糕）一词来源于 croustade（脆皮酥盒）或者 crust（外皮），根据《牛津英语词典》的解释，蛋奶糕"曾是一种包含肉片或者水果片的开口派，上面覆盖着肉汤或者牛奶，用鸡蛋增稠，加糖增甜，并用香料等材料进行调味"。今天的 dariole 是一种奶油小蛋糕，但是在过去这是一种"把肉、香草、香料切碎混合在一起作为馅料"的小型派。甚至连 rissole（炸肉饼）曾经也是"一种切碎的派"，通常通过油炸制成（源自法语 rissoler，意为炸至金褐色）。

这些从派衍生出来的词非常幸运地得以流传至今，而更多的相关词汇现今已不复存在。你上一次在菜单上看到chewet（一种小型的圆派，把肉或者鱼切碎混合香料和水果制成，"比西葫芦派要高"）、dowlet（一种小型的派，通常装饰着精美的小花边）、herbelade/hebolace（一种猪肉末与香草混合制成的派）、talemouse（一种芝士蛋糕，形状有时为三角形）或者vaunt（一种水果派）还是什么时候？以上名称以及更多其他同类食品的名称，曾经对于烘焙师来说都耳熟能详。在这里我们可以得出的唯一结论是，名称不复存在，说明这些派已然消失。

派消逝了吗？

派曾是英国人的"肉和土豆"[①]，而在19世纪，随着真正的土豆产量增加，派的重要性开始下降。进入20世纪之后，社会变革使派进一步衰落。在英国爱德华时代，也就是各阶层人士离开故土、入伍参加第一次世界大战之前，大型派在英国本土的庄园里度过了最后的光辉岁月。普通的家常派则坚持了更长时间，在二战期间依然很常见。后来，越来越多的家庭主妇开始参加工作，做饭的时间越来越少，于是家常派也逐渐没落。从此之后，人们对于时间的紧迫感越来越强烈，自然也就首先抛弃了生活中最耗时且最麻烦的烹饪环节。与此同时，营养"警察"的宣传也

---

① 原文为 meat and potatoes，意为最基本、最必要的部分，这里指必需食物。——编注

势不可当地向人们扑面而来。从各个角度来看，家庭自制派所遭遇的困境已经至少持续了一百年。派的存活无疑受到了威胁。

然而奇怪的是，我们仍然没有彻底放弃派这种食物。尽管派正在逐渐从餐桌上消失，但是我们仍然会怀着某种莫名的热情坚持派的"理念"。为什么会这样呢？派究竟是一种怎样的存在？

派的普遍吸引力

除了走上战场之外，一个男孩也有其他途径成为英雄。他可以在派不够分的时候，说自己不喜欢吃派。

——埃德加·沃森·豪

毫无疑问，对于 19 世纪的厨师和食谱作家来说，派有一种令人难以把握的特质，这种特质使派受到普通推崇。而其他食物，诸如蛋糕、炖菜或者汤，并不具备这种特质。简单的几个引证就足以说明这个问题。

1806 年，朗德尔夫人在她的畅销书《居家烹饪新法》中展开了"对美味的派的观察"。她自信地说道："只要操作得当，没有什么菜品会比美味的派更受人喜爱。"著名厨师亚力克西斯·索耶在他的作品《给人们的一先令烹饪法》（1860 年）中，提到了派在维多利亚时代日常生活中的重要性：

从小到大，我们都吃派。不论男孩女孩，我们都吃派。从中年到老年，我们都吃派。事实上，在英国，我们或许可以认为派是人生旅途的最佳伴侣之一。派没有离开我们，是我们离开了派。我们的子孙将会像我们一样爱吃派。因此，我们应该学会如何做派，以及如何把派做好！相信我，我没有开玩笑：如果把周日一天之内整个伦敦做坏了的派沿着铁路排成一排，要坐一个小时的专列才能把这些烹饪受害者看个遍。

社会新闻记者查尔斯·曼比·史密斯在自己的《伦敦生活奇闻》（1853年）一书中对派进行了更高的评价。他把派称为"人类的伟大发现，受到所有文明饮食者的普遍好评"。

当然，任何一种观点都会有人持反对意见，但少数消化不良的坏脾气和禁欲主义者不认可派的吸引力，只起到了再次证明这一法则的作用。安布罗斯·比耶尔斯在他的《魔鬼词典》中把派定义为"一个名叫'消化不良'的死神的先遣人员"。但这里或许仅仅是他在说明个人的健康问题。19世纪的营养大师西尔维斯特·格雷厄姆曾经成功说服一大批美国民众相信，所有社会问题的解决办法以及获得上天恩赐的担保都在于日常严格进行冷水浴、清淡饮食以及抑制性欲，而且霍乱是由鸡肉派引发的。如果可怜的格雷厄姆

一个穿戴着烘焙师帽子和围裙的小孩在吃派。

能多活几十年，看到霍乱其实是来自水污染的证据，那么他很有可能会允许自己偶尔吃个派来弥补自己的性压抑。

中世纪的野餐

《草地上的午餐》（1865—1866，克劳德·莫奈绘）中的一部分细节

派的作用

　　派最初有三个非常实用的功能：在烘焙、携带和保存食物的过程中，它都可以充当容器的角色。当代社会已经可以通过其他更有效的方法实现上述功能，但对于厨师和食客来说，派仍然是一种特别实用的食物。在纯粹的多功能属性方面，没有任何一种食物能够超越派。面包非常实用，汤也还算实用，但正是由于多功能性，派拥有着它们所不具备的用途。派可以冷着吃，也可以热着吃，可以作为一日三餐中的任何一道菜，还可以在野餐或者旅途中食用，而且特别适合需要用手拿着吃的场合。它们可以做得简朴，也可以做得豪华；可以是日常简餐，也可以是特殊盛宴；既是一种简单的食物，又是一种有力的象征。派可以因环境、想象力和良心而更换各种各样的馅料，派的外皮也是展示厨师艺术造诣与烹饪技巧的绝佳机会。

　　派具有一种突出的历史特征：形态独立，可以拿在手里

中世纪的便携式派烤炉

第一次世界大战期间"联合战时工作运动"的海报

吃，不需要刀叉、碗碟或者桌布、餐巾。这种便捷至极的
特点可能构成了派在英国受到广泛喜爱的原因之一。希拉·哈
钦斯在她的作品《英国食谱》（1967年）中写道："热衷于斗
鸡和板球且有着良好体育风尚的英国贵族们早期便形成了
一种不干扰手头活动的饮食体系。"哈钦斯还用了三明治伯
爵①的故事来举例：1762年的一天晚上，三明治伯爵不愿离
开牌桌，于是吩咐人给他在两片面包中间夹一片牛肉，由
此便发明了三明治，伯爵本人的名字也永远为世人所铭记。
这是一个有趣的传说，但我不太理解：那个时代和当时的英
国已经以派著称，伯爵为什么不叫人端来晚餐剩下的派呢？

　　派可以重复加热的特性对于厨师来说通常是一种优点，
但对于食客来说往往是一种缺点。14世纪的乔叟②非常了

① 约翰·孟塔古，第四代三明治伯爵（John Montagu, 4th Earl of Sandwich,
1718—1792），英国政治家、军人，曾三任第一海军大臣。——译注
② 杰弗里·乔叟（Geoffrey Chaucer，约1343—1400），英国小说家、诗
人。主要作品包括《坎特伯雷故事集》《公爵夫人之书》《声誉之宫》等。
——译注

解这一点，于是在《坎特伯雷故事集》中进行了精彩的阐述。在这部作品的有关章节中，一名厨师因为向朝圣者提供了重新加热的派（厨师事先吸走了派里的一些肉汁），而被要求讲述自己的故事作为弥补：

> 讲吧，罗杰，讲一个好故事，
>
> 因为你夺走了很多派的血液，
>
> 你还卖出了很多瓶多佛杰克，
>
> 加热后冷却，重复了两次。

这里的"多佛杰克"是一种用从其他酒瓶里收集来的酒渣制成的葡萄酒，换个瓶塞后当作新酒出售。乔叟在这里用这种酒比喻重复加热的派，并且把厨师写成了一个腿上长着脓疮的脏兮兮的人，使厨师的行为更加令人反感。

索耶把派形容为"人生旅途的最佳伴侣之一"，这句话是对食客说的，但同样有可能是对烹饪领域的同行们说的。派对于那些餐饮供应者和厨师来说一直非常有用，尤其是在那种必须高效地为大批食客提供食物的场合。现代背景下的大批量供餐通常发生在足球比赛等体育赛事场合，然而这方面最早的专家其实是军队。1891 年 7 月 11 日，德国皇帝访问英国，在温布尔顿检阅了军队。当时的邮局志愿者由军需官迪克森主管。迪克森发明了一套高效且简单得惊人的系统，给记者留下了深刻印象。记者描述道：

每个排队路过的人一手拿起一大杯酒，另一只手领一份猪肉派，毫无停顿，然后走到一百码开外的地方享用午餐。结果是，就在 7 分半钟的时间内约有 800 人领到了餐食。

军需官迪克森显然超越了当时的时代。即使在今天速度最快的快餐店里，7 分半钟供应 800 名顾客，同样也是一项非常惊人的成就。

生理之因

从生物学角度来说，有个无法逃避的事实——节食者面临的绝望无处不在。出于生存目的，我们生来就非常渴望营养密度高的食物（比如派），会在获取这一类食物的时候尽量多吃一些，把食物高效转化为体脂肪，为应对下一次饥荒做好准备。原始人类不需要对纤维和抗氧化物产生渴求，因为当时的生活环境肮脏又野蛮，人类的平均寿命很短，纤维和抗氧化物之类的物质发挥不了作用。那么，在没有营养学家和饮食专家提供帮助的情况下，早期智人是如何判断哪些食物营养密度较高的呢？

人类面对食物的时候，首先会被气味所吸引。我们的鼻子可以探测到周围化学环境中的挥发性成分，温度上升的时候这些挥发性成分也会增加。因此我们可以猜到，那些经过烹饪的高脂肪、高蛋白食物的味道对于我们这个物种来说最具有吸引力。事实也确实如此。从简单的生物学角度来看，在饮食专家出现之前，主要由我们的鼻子来告诉我们最好应该选择哪种食物：闻起来不错的食物在多数情况下必然好吃。英国首次出现增值税（针对产品和服务征税）的时候，烘焙师们成功为热肉派争取到了免税待遇。他们争辩说，店铺之所以给派加热，主要目的在于制造诱人的香味，并没有提供任何特殊服务；即使是让人最没食欲的派，在热的状态下闻起来也会很香。

在很长一段时间里，西方人普遍认为味蕾只能分辨四种基本的味道——甜、咸、酸、苦。现如今，大多数科学家已经接受了日本人几百年前的观点：世界上的第五种味道是"鲜味"。"鲜味"在概念上就是"令人愉悦的味道"，比如那些陈年奶酪、烤肉以及鱼露所包含的味道。这种味道主要来自于谷氨酸。谷氨酸是一种氨基酸，而氨基酸又是蛋白质的组成成分。从生物学角度来讲，人体具有探测蛋白质的机制是有意义的。甜食可以为我们提供热量，咸味食物可以维持人体化学环境的平衡；酸味和苦味在自然环境下通常代表有害，因此这两种味道可以使我们保持警惕；鲜味则可以唤醒人体对蛋白质的感应。母乳有鲜味，其谷氨酸含量相当于牛奶的十倍。肉派同样有鲜味。

我们在进食的时候还会经历第三个步骤：从触觉上感受食物。食物的质地以及浓郁程度，即我们所谓的"口感"。如果人类对美食的享受只有嗅觉和味觉（进入文明社会之后或许又增加了视觉），那么我们完全可以把派打成浆倒进杯子里，在杯子上插一个小小的纸伞装饰一下，照常享用。派的好坏取决于油酥面皮，一份质量上乘的派所带来的乐趣之一，恰恰就在于酥脆的油酥面皮与各种馅料在口感上形成的鲜明对比。对于一份完美的派来说，各个组成部分单独拎出来都应该是完美的。油酥面皮应该有着轻薄、松脆、黄油般的口感，馅料则应该具有浓郁、油润、软嫩、黏糊、爽脆等口感，而派整体的口感要更甚于各部分相加之和。

因此，难道派对于人类感官来说不是一种相当完美的食物吗？还有任何其他的单个菜品能在这方面与之匹敌吗？

热量之外

人们对于派的喜爱具有坚实的生物学基础，但仅靠营养密度这一点已经不足以满足现代人的需求，否则我们完全可以像吃派那样去享用鲸油。人类是群居动物，通常不会独自觅食和进食，因此我们会在情感方面把食物与人、事件以及环境关联起来。最终，一种食物就被赋予了意义，使得人类学家会提出类似这样的问题："派是否具有什么含义？"

气味是一种强大的情感触发器，因为人类会把气味与他们感知的社会情境、情感情境联系在一起。芝加哥的嗅觉和味觉治疗与研究基金会针对"嗅觉诱发记忆"展开了一些非常有趣的研究。当然，研究的受试者均来自芝加哥，而他们的祖先很有可能来自派的故乡——欧洲，所以研究

布面油画《船长和他的船员》（又名《餐前祈祷》，1881 年，阿瑟·休斯绘）

结果可能无法适用于其他人群。不过这项研究的结果非常有趣：最能引起怀旧情绪的气味来自烘焙类食品，比如面包、蛋糕等等，紧随其后的是含肉菜肴，如培根意面和肉丸意面。在我看来，肉派是以上两者的结合，如果它作为一个单独的类别出现在选项中，肯定可以轻松夺冠。

当然，这对于许多不从事科学研究的人来说也并非什么新鲜事。房地产经纪人早就深谙其道：他们会建议房产卖家在房子里放一壶刚刚煮好的咖啡，这样潜在买家来看房子的时候，房子闻起来就会有家的味道。普鲁斯特[1]比任何人都更好地解释了这一点——他曾经描写道，黄油小蛋糕

---

① 马塞尔·普鲁斯特（Marcel Proust，1871—1922），20 世纪法国乃至世界文学史上最伟大的小说家之一，意识流文学大师。代表作品：《追忆似水年华》。——译注

蘸上清新的柠檬花茶之后产生的香气，使回忆如潮水一般涌上心头。

克雷格·克雷本[1]曾说："我明白，没有什么可以比得上一个人童年以及少年时期的食物所具有的吸引力。"这些食物可以唤起我们共有的关于家的记忆。在我们的文化意识里，这一类食物具有抚慰效果，与母亲以及祖国密不可分。尽管派已不再经常出现在我们日常的餐桌上，它们也仍然在我们的各种记忆中占据举足轻重的地位——感恩节、圣诞节、野餐、笨笨的"老阿姨梅布尔"[2]，以及和爸爸一起去踢足球，这些都是关于派的记忆。如果说派的香气"售卖"的不仅仅是产热营养素的吸引力，而且还包含了关于派的过往记忆，那么对于制作派的厨师和食用派的顾客来说，都实属幸事。

派，特别是那种家庭自制的派，还具有另外一种含义，这一含义概括了派受到普遍推崇的原因。用玛格丽特·富尔顿[3]的话来说就是："派一直被誉为一种款待，也是一个体贴的厨师的象征。"对方出于对你的爱而大费周折，他们没有把炖菜和一大块面包随意摆在盘子里，而是为你精心制作了一个小小的油酥面皮礼盒。这种心意才是关键所在，对吗？

---

① 克雷格·克雷本（Craig Claiborne，1920—2000），美国著名美食评论家、美食记者。——译注
② 20世纪90年代英国儿童电视连续剧《出来》（Come Outside）中的主要人物角色。——编注
③ 玛格丽特·富尔顿（Margaret Fulton，1924—2019），澳大利亚美食烹饪作家、记者和评论家，出生于苏格兰。——编注

［第三章］

派的设计感

美的艺术总共包含五种形式：绘画、雕塑、诗
歌、音乐、建筑。而建筑的主要分支就是糕点。

　　　　　——安托南·卡雷姆（马里-安托万·卡雷姆）[1]

　　安托南·卡雷姆曾被誉为"王之厨师，厨师之王"，而
讽刺的是，如果他可以选择，他很有可能会去学习建筑。
卡雷姆在童年时期因家境贫困而遭到父母遗弃，于是找了
一份厨房打杂的工作谋生，就此开启了他的烹饪生涯。卡
雷姆的余生都在厨房里工作，但是在内心深处，他仍然是
一名建筑师。华美至极的"装置糕点"是他的标志，他就

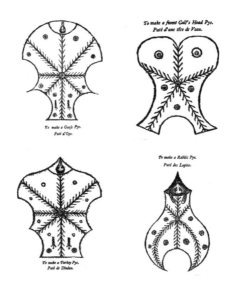

派的各种形状。图片出自约翰·撒克的《烹饪的艺术》（1758 年）。

①　安托南·卡雷姆（Antonin Carême，1784—1833），法国现代美食之父，
法国烹饪史上首位将烹饪进行分门别类做了系统化整理的厨艺大师，同
时也是国际首位明星主厨。——译注

在设计与制作它们的同时表达着自己的满腔热情。

也许卡雷姆的话是有道理的。毕竟一份派恰如一座建筑，需要"施工建造"。同样像建筑的是，单一尺寸与样式的派并不适配所有的馅料和场合。20世纪早期的建筑师或许会认为是他们首次提出了"形式服从于功能"的理念，但对于糕点师而言，这项原则他们已遵循了数百年之久。一份包含了一整块鹿肉的派是为了长时间保存鹿肉而制成，在设计上与基督教大斋节期间允许食用的精致的杏仁蛋奶糕相比，完全是两个不同的世界；与那些由活力四射、衣着暴露的年轻女人做出的与烹饪毫不相关的食物相比，更是两个完全不同的宇宙。环境的差别同样可以决定设计的差别：一个派足够坚固，才能胜任长时间的航海；足够优雅，才能被作为礼物赠予他人；足够别致，才能愉悦食客；或者足够威风，才能展现权力或达到宣传效果。

随着17世纪的发展，越来越多的烹饪书籍开始面向非专业人士，降低了知识门槛，这对于家庭主妇来说是一件好事；对于我们其他人来说同样实属幸事，因为这些书籍让我们对油酥面皮和派的制作有了更加深入的了解。杰维斯·马卡姆在他的作品《英国主妇》（1615年）中煞费苦心地罗列出了适用于不同派的不同种类的油酥面皮。

英国的家庭主妇必须懂得如何制作油酥面皮，

如何烤制不同种类的肉，哪种油酥面皮适合哪种肉，

以及如何调制油酥面皮面团，比如马鹿肉、野猪肉、

腌培根、天鹅肉、麋鹿肉、海豚肉等等，都需要不同的面团处理方式。再比如立式派，这种派必须能够长时间保存，应该烤制出微湿、粗糙且能长时间保存的厚皮，因此黑麦面团最适合制作这种皮。包裹火鸡肉、公鸡肉、野鸡肉、鹧鸪肉、牛犊肉、孔雀肉、羊羔肉以及各种水禽肉的派通常可以分多次食用，不过也不能放置太多天，烤制这种派就需要较厚的优质白皮，因此应选用小麦来制作。含有鸡肉、牛腿肉、橄榄、土豆、榅桲[①]、黇鹿肉等食材的派一般需要趁热吃，且外皮非常松脆、纤薄，要做出这种外皮，你最好在揉面之前把小麦粉放在烤炉里稍微烤一下。

马卡姆在作品中提到，"派必须能够长时间保存"，这是因为在他所处的时代，"长时间保存"对于派来说是最重要的要求之一。在制冷技术和罐装技术出现之前，保存肉类的方法仅包括干燥、烟熏、盐渍，或者用"厚且粗糙"的外皮包裹起来。可以使派长时间保存的油酥面皮通常由黑麦粉制成，厚度可达几英寸，会烤至非常坚硬，且这种油酥面皮"是用来保存里面的食物，不是用来吃的"。待派烘烤完毕从烤炉取出的时候，需要通过派皮顶面的一个洞注入融化的油脂以排出空气，这样就可以保存馅料。一旦

---

① 榅桲（quinces），也称木梨，一种外表类似梨子的水果。——译注

把派切开，这种密封状态就遭到了破坏，里面的食材就会迅速变质；有一种古老的迷信观念认为从整个派里只取走一小块是不吉利的行为，这或许就是此迷信的由来吧。

古老烹饪书中的一些指导说明有可能是为了最大限度地保存派，对于今天的我们来说意义不大。15世纪的一份手稿中有一个很奇怪的警告："不要让番红花靠近派的边缘，否则派永远都合不上"。在当时，无论制作油酥糕点还是派，番红花都是一种常用原料，所以这一说明非常令人费解。16世纪，德国的萨比娜·韦尔斯林写了一本烹饪书，书中说明了如何把派的顶面与底部油酥面皮结合起来——首先需要用手指把油酥面皮捏合在一起，对此我们现代人非常熟悉。但随后她建议道：

> 留一个小洞。再看油酥面皮是否已经充分捏合，避免裂开。向这个小洞里吹气，酥皮顶面就会鼓起来。这时迅速把小洞捏合。

这种方法是只为做一个美观的圆顶，还是同样为了让外皮和多汁的内馅分离呢？外皮确实必须保持非常干燥，否则便失去了保存的功能。威廉·萨蒙在他的《家庭词典，或家居伴侣》（1695年）中给出了使用黄油密封野猪派的食谱，这种方法"可以使派在不太潮湿的环境下保存一整年"。在缺乏制冷技术的情况下把一个派保存一整年，这个想法在今天看来简直恐怖，然而这种做法在当时极为常

野猪派

见，可见这种保存方法在当时的多数情况下并没有对食客的健康造成严重危害。

厚实的外皮同样形成了一种坚固的容器，使派可以经受住远距离的陆路和海路运送。19世纪，英国人常常用火车将那些巨大的"约克郡圣诞派"运送至伦敦，而这与数世纪以来那些尽心尽力的母亲和妻子所做出的努力相比，简直不值一提：她们总是想方设法地将派送至在远方上大学或打仗的儿子和丈夫手中。

布里利亚娜·哈莉夫人是来自英国赫里福德郡布兰普顿·布莱恩村的一位母亲，1638年至1639年间，她曾经定期托送货人为她在牛津读大学的儿子爱德华送去派。送货人从赫里福德郡到牛津市可能要花好几天时间，而在几年的大学时光里，爱德华并没有出现健康方面的严重问题，由此可见哈莉夫人做的派非常坚固且密封性很好。5月10日，哈莉夫人为爱德华寄出了一个小山羊肉派：

我给你做了个派送过去，是小山羊肉派。我觉
得你在牛津平时肯定吃不到这种肉。这个派一半是
一种口味，另一半是另一种口味。

邮政服务一出现，人们便开始通过寄件的方式向各地
邮寄派。1884 年，邮局给出了关于如何在圣诞节期间寄派
的温馨提示：

如需通过邮政包裹邮寄圣诞节礼物，如冬青、
槲寄生或其他类型的装饰品，禽肉、野味、布丁、
肉馅派或任何其他各种类型的油酥糕点，以及糖果、
苹果、玩具、精美物品等，邮寄的同时请仔细进行
包装，以避免物品受到损坏。

极为可靠的邮局甚至成功避免了这些好东西在邮寄途
中失窃。一名在南非参加布尔战争①的英国士兵大受感动，
特意致信《泰晤士报》：

我想赞扬前线的邮政包裹服务……虽然我一
直在请求寄出包裹的人不要在包裹上粘贴巧克力、
野味派、鸡肉或者舌头之类的标签……但我确定

① 布尔战争一般指第二次布尔战争，即 1899 年 10 月 11 日至 1902 年 5
月 31 日，英国同荷兰移民后代阿非利卡人（布尔人）建立的德兰士瓦共
和国与奥兰治自由邦为争夺南非领土和资源爆发的一场战争。——译注

我一个包裹也没丢。

派的艺术

那些曾经宣称"形式服从于功能"的建筑师还曾经主张，所有的装饰都有害无益。很明显，糕点师从来没有想过要遵循后面这条理念。糕点是一种极佳的艺术媒介，"奇妙糕点艺术"的一代又一代从业者都饱含热情地投入到了创作之中。

工业革命带来的技术进步使人们制作出了精美绝伦的派模具。不过早在此之前，很多糕点师就已经利用烫面酥皮的可塑性，为派制作出了各种巧妙的造型，比如鸟形、城堡形、鱼形，或者那个藏于烘焙师内心深处的艺术家所希望做出的任何其他艺术造型。城堡形状的派一直广受欢迎。（也许早期的糕点师都是不得志的建筑师？）《烹饪之法》这本书里有一份"城堡派"的食谱，这是一种城堡造型的派，四座齿状塔楼围绕着中心的庭院，每座塔楼都有不同的馅料和颜色。18世纪末，詹姆斯·伍德福德牧师在日记中记录了一顿餐食，其中包括"用塔楼状面团包裹着的牛排挞"。

有时候大尺寸的派在端上桌之前，厨师会去掉原本顶部的油酥面皮，替换成一块独立烤制且装饰成盾徽等造型的油酥面皮；或者去掉顶部油酥面皮之后，使用香草、鲜花或者从油酥面皮和彩色果冻上裁下的图案在馅料顶部进行装饰，想做得多艳丽都行。根据 1658 年的一份牛排派食谱，去掉顶部油酥面皮之后，可以将煎过的鼠尾草叶子直立插在油酥面皮里侧，想必这个造型的派看起来就像一座小型的室内花园。

17 世纪的烘焙师会使用裁成野兽、鸟类、武器、绳结、鲜花等各种造型的纸模板制作油酥面皮，把派端上桌之前用事先烤制好的这些不同造型的油酥面皮装饰派的顶部。部分造型还会涉及针线活：有一种派叫作"刺绣派"，之所以有这样的名字，是因为这种派在制作过程中使用了詹姆斯一世时期流行的立体浮雕刺绣，给派添加了复杂的油酥面皮绳结、鲜花、纹章图案等装饰。

PATE-CHAUDS, OR RAISED PIES FOR ENTREES,

COMPRISING

Pâté-Chaud of Young Rabbits with fine-herbs.	*Pâté-Chaud* of Young Partridges, à la Chasseur.
" à la Sauce Poivrade.	" of Ox Palates, à l'Italienne.
" of Leverets with Truffles, à la Périgueux.	" of Quails, à la Financière.
" of ditto à la Financière.	" of Larks boned, à l'Essence.
" of Godiveau à la Ciboulette.	" of Snipes, à la Bordelaise.

Pâté-Chaud Cases.

752. PATE-CHAUD* OF YOUNG RABBITS, WITH FINE-HERBS.

装饰繁复的立式派，这些派出自维多利亚女王的主厨查尔斯·埃尔姆·弗朗卡泰里之手。图片出自《现代厨师》（1860 年）。

过去的那些"油酥面皮艺术家"也会使用色彩对油酥面皮进行装饰，比如把糖和玫瑰水混合起来给顶面油酥面皮添加糖霜，使用番红花、蛋黄或者真金为顶面油酥面皮涂上一层金色，或者使用各种着色剂给顶面油酥面皮上色。从中世纪开始，很多书籍中都出现了着色剂配方，其中的一些配方看起来相当令人担忧。《寡妇的宝藏》（1586年）这本书给出了祖母绿色着色剂的配方，原料包括醋酸铜、氧化铅和水银（金属汞），还要使用小孩的尿液把这些原料混合起来。书中还有一种金色配方，需要将番红花、雌黄（三硫化二砷）连同野兔或梭子鱼的胆汁一同捣碎，然后装入小瓶放进粪堆里埋五天。某些工艺的消失对人类来说或许是一件好事。

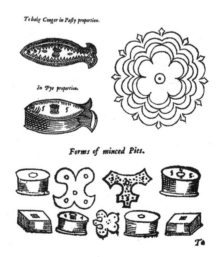

派的各种形状轮廓。图片出自《卓越厨师》（1660年），罗伯特·梅著。右上方的图形为"新娘派"的造型（参见第71页）。

曾经的派通常尺寸很大，需要能够喂饱很多人。在当时，恰当地上派也是一种艺术。对于绅士来说，优雅且正确地切分不同种类的带骨大肉块是一项基本技能，也是一项很荣幸的任务。绅士还需要了解正确的术语，如"肢解苍鹭，鲟鱼切片，分割天鹅，剖开孔雀"或者"沿边切分这个派"，而且他们也应该知道，如果派需要趁热吃，就在顶部切开；如果需要凉着吃，就从中间切开。

当然，如果不想表现得粗鲁，那就要知道怎样正确地吃派。好在几个世纪以来一直流传着一些礼仪手册，讲述了正确的吃派方式。1609 年出版的一本儿童礼仪书提醒孩子们："如果有人给了你一块派或者水果挞，你应该用碟子或木盘子接过来，不要用手接。"1853 年，女士们被建议道：

> 直接使用叉子吃派是一种时尚过头的矫揉造作之举，而且非常不方便，吃相尴尬。应该先使用刀叉把派切成小块，然后再用右手拿叉子吃……在公共餐桌上，女士绝不要主动为其他人拌沙拉，也不要亲自切派或者给周围的人分派。这一类事情只应该由绅士或者用人去做。

社会生活中的派

从理论上来说，现代社会里所有人都可以赚到钱。因此现代人很难理解，在过去的时代财富完全与社会阶层相捆绑，社会阶层决定了你能吃到哪些东西，甚至决定了你的派的油酥面皮类型。17世纪与18世纪的一些农业与家务手册明确说明，主人及其家人的派皮由最精细的小麦面粉制成，而仆人的派皮由二次碾磨的小麦或者大麦制成，也可以由黑麦或者黑麦与小麦的混合面粉制成。

维多利亚时代的鸽子派示意图。注意图上的鸽子脚。

根据有关界定，野味派和很多种类的鱼派只能由那些拥有土地和狩猎权的人享用。在维多利亚时代的英国，穷人和工人阶级如果有机会吃到羊肉派，也大多是年龄较大的羊的肉；他们吃的牛肉派使用的牛肉则大多来自年老的乳牛或役用牛。1857年，一份英国报纸刊登了一篇关于著名的德比赛马会的文章，这篇文章对不同阶级的食物划分进行了很好的总结：

德比赛值得一看。我不知道还有什么能和德比赛一样充分展示英国的整体现状。会场上聚集着各阶层样本——最高贵的贵族与最底层的民众。主教们乘着马车，带着一篮又一篮的野味派和香槟；小贩则满载面包、奶酪和啤酒，驱着驴车。

Mayonnaise of Salmon

Raised Pie.

Lobster Salad.

Cherry Tartlets.

Game Pie.

Fancy Pastry.

Open Tart.

Tomato and Cucumber Salad.

Ratafia Pudding.

Pigeon Pie.

Meat Pie.

Supper Dishes.

"晚餐菜肴"。图片出自《比顿夫人的日常烹饪》。

在英国，鸽子派是一种供上层人士食用的高端美食，因为只有那些富有的地主才能拥有鸽舍，在冬天吃上新鲜的鸽子肉。而美国最初的情况刚好相反。在 19 世纪初的美国，成群结队的候鸽简直遮天蔽日，可以覆盖几英里的范围。如果晚餐想吃上鸽子派，只要有枪，任何人都可以轻而易举地随意捕杀候鸽。捕猎者们的成功导致候鸽在 1914 年灭绝，不过灭绝之前的一小段时间里，稀缺的候鸽一度成为有钱人在餐厅里享用的昂贵美食。

在 18 世纪的餐桌上，派的世界也存在着自己的等级秩序。今天我们的上菜方式为"俄式"，每一道菜品都按照先后顺序摆到客人面前。但当时的上菜方式为"法式"，两道或两道以上的菜同时摆到餐桌上，每一道菜又由多种菜品组成，按照精确的几何布局或者对称美学进行摆放。派摆在餐桌上具有突出的视觉效果，因此在"法式"上菜方式中具有非常重要的地位。大型的派一般摆在餐桌最中央，较小的派摆在边缘或者角落里（其实有些小型派就叫作"角落派"）。在这段历史时期，派的馅料种类之多令人叹为观止。

馅 料

好的苹果派是家庭幸福的重要组成部分。

——简·奥斯汀

《牛津英语词典》针对 stuff[①]（东西、玩意儿）这个词列出的其中一条定义是"用来填充派的材料"。词典在这里并没有尝试列举材料种类，而且也无法进行列举。因为我们能想象到的各种东西，高端也好，邪恶也罢，都会在某个时间段以某种特定方式成为派的馅料。

我们已经知道，派最初的馅料是一大块肉。到了 14 世纪，馅料的范围已经扩大到包含水果、蛋奶酱以及经过调

《一桌甜点》（1640 年，杨·戴维茨·德海姆绘）中的部分细节

---

① 本章节英文名为 Filling Stuff，故此处提到 stuff 这个词。——编注

味的肉馅等食材。今天的我们如果看到 14 世纪的食谱，会惊讶地发现很多咸味派都含糖。甜味食物和咸味食物之间的区别是一个相对现代的概念，中世纪的厨师并不知道这种区分。当时的糖是一种价格不菲的进口原料，厨师把糖当作香料使用。16 世纪中叶，英国本土出现了一批制糖厂，糖的价格开始下跌。随着 17 世纪东印度贸易的发展，糖变得更便宜了。

到了 17 世纪中叶，人们开始大量使用糖，不再像使用香料那样少量添加。由此也可见当时的人们很喜欢甜味。在 17 世纪的著名畅销烹饪书《卓越厨师》中，罗伯特·梅收录了这份迷人的鲱鱼派食谱：

如何制作鲱鱼肉馅派

把腌过的鲱鱼湿水之后用手捏碎，把鱼肉和鱼皮分离，去掉整块鱼皮，再把鱼肉放在盘子里。准备一磅①杏仁酱，把鲱鱼捣成肉馅，然后在肉馅里添加杏仁酱、两个鱼白、五六个海枣、适量白面包屑、糖、加强白葡萄酒、玫瑰水以及藏红花，把混合起来的馅料搅拌浓稠之后就可以准备装进外皮。在派的底部放入黄油，铺上馅料，然后放上海枣、醋栗、葡萄干、伏牛花以及黄油，把外皮封起来进行烘烤。烘烤的时候在外皮表面涂上黄油、酸果汁和糖。

---

① 1 磅约等于 453.59 克。——译注

到了 18 世纪，咸味菜品和甜味菜品区分得越来越明显，当时的很多烹饪书也提供了两种不同口味肉派的食谱，以满足不同的口味偏好。以下食谱出自汉娜·格拉斯的《简单明了的烹饪艺术》（1747 年），食谱后面又给出了咸味版本。

制作美味的甜小羊肉派或小牛肉派

使用盐、胡椒、丁香、肉豆蔻衣、肉豆蔻按你的口味给小羊肉调味，搅拌均匀。把小羊肉或者小牛肉切成小片。准备一块优质的千层酥皮，把酥皮放在盘子里，然后把肉铺进酥皮里。在肉上撒一些洗净的去核麝香葡萄干、黑科林斯葡萄干以及适量的糖。接下来铺上一些甜味碎肉丸，夏天可以加一些煮过的洋蓟，冬天可以加一些烫过的葡萄。把西班牙土豆切块煮熟，要连同香橼蜜饯、橙子蜜饯、柠檬皮、三四片大块的肉豆蔻衣一起煮。给馅料顶部铺上黄油，把酥皮封起来，开始烘烤。现在开始准备烘烤完毕之后需要用到的酒汤：一品脱①白葡萄酒外加三个蛋黄，在火上顺着一个方向搅拌，搅至浓稠就从火上取下来，加入足够的糖增甜，再挤一点柠檬汁。把酒汤趁热倒进派里面，再把派封起来，就可以趁热端上桌了。

---

① 品脱，容量单位。1 英制品脱约等于 568.26 毫升。——译注

有着独特造型的小派，来自法国南部的佩兹纳斯镇。

肉馅、香料、糖、果干——这个食谱充满了中世纪的气息，如今我们也仍然可以在圣诞百果派里感受到这种气息。法国朗格多克地区有一个小镇名叫佩兹纳斯，这里的中世纪气息尤为浓烈。当地有一种特产叫作"佩兹纳斯小派"，这是一种非常小的派，一口就能吃掉，形状像花盆，馅料为甜小羊肉。1786 年，极具传奇色彩的印度克莱夫勋爵曾经在佩兹纳斯住了几个月。尽管身体欠佳，克莱夫勋爵仍然终日忙于社交。当地人对勋爵家中的派很感兴趣，于是说服勋爵的厨师提供了这种派的食谱。这种派成为佩兹纳斯的当地特色菜之后就再也没有发生过变化。它像是烹饪界的一粒时间胶囊，经久不衰，让我们得以瞥见那个时代的风采。加拿大的新斯科舍省也有一种穿越了时间的美食——没有猪肉的"布雷顿角猪肉派"。这种派里现在放的是椰枣，大概取代了过去的猪肉馅。

鱼派

在可以选择的情况下，鱼肉可以很好地替代禽兽肉，同样特别适合做馅料。在历史上的很多阶段，教会都规定了不吃肉的斋戒日，而且这种规定通常有成文法律的支持。在某些时期，斋戒日天数几乎可以占到全年总天数的一半。在斋戒期间，尤其是一些重大场合，鱼派扮演了非常重要的角色。

禁止吃禽兽肉的原因有很多。在古代，人们认为禽兽肉可以"点燃激情"，使人从更高层次的思想境界中分心，而鱼（确切地说是水生动物，这里包括了鲸鱼和海豚）则可以让人冷静下来。当时的人们认为，自然界中的所有生物都可以把自身的特征与习性传播给捕食者，而鱼没有表

中世纪厨房里骚乱的人群和几种典型的派和馅饼

现出明显的性行为，因此非常适合用于宗教斋戒日。后来，许多非宗教因素也巩固了斋戒期间可以吃鱼的规定。比如在农业方面，吃鱼可以给国家节省大量的禽兽肉；在经济方面，捕鱼业受到了国家的支持；在政治方面，捕鱼业的发展可以为海军和勘探航行提供专业知识和人力资源储备。

当然，如果有人非常渴望肉，那么也完全有可能对成文的斋戒规定阳奉阴违。有些为斋戒日制作的丰盛的鱼类菜肴和"斋素"毫无关系。《厨艺大全》（1658年）的作者给出了一份鲤鱼派的食谱（派里塞满了鲤鱼血和肥嫩的鳗鱼肉），并在结尾评论道："这是供教皇食用的肉。"以下这份食谱出自1702年的一本法国烹饪书，这种派绝对是新奇鱼派的终极版本，但它也确实非常适合用来应对斋戒日：

使用鲤鱼卵和舌头制作平底锅派

把鲤鱼卵和舌头整齐地铺在一块油酥面皮上，放入平底锅。使用胡椒、盐、肉豆蔻、香草末、大葱、羊肚菌、常见蘑菇、松露以及甜黄油调味。用同一面团做成的油酥面皮顶将所有原料封上，用小火把派烤熟。最后配上柠檬汁端上餐桌。

在基督教大斋节的40天时间里，鱼肉也遭到了禁食。在某些历史时期，乳制品和鸡蛋同样被禁止食用。于是厨师们开始发挥创意，使用杏仁奶、大米、坚果、水果等食材作为替代品来制作派——他们的尝试使派的食谱库得到了扩充。

甜　派

到目前为止，我们提到的大多都是咸味派。如果你很喜欢吃甜食，而且一直把派当作一种甜品，那么上文的内容可能无法满足你的胃口。接下来，我们一起进入甜派时间。

16世纪，糖的价格下跌，市面上出现了更多精致的油酥面皮造型，水果派也开始盛行起来。在此之前的派其实也会用到水果，而且用得很多，只不过很少把水果作为主料。最早的水果派仍然被称为 bake-metes［meat（肉）在过去可以指代各种固态食物］，但这种水果派并不是我们概念中的水果派（这种水果派外皮很厚，但与如今的水果派之间的区别不止于此）。有一份中世纪的"烤梨肉"食谱要求在梨片之间加入"小块骨髓"（gobbets of marrow，此处的marrow 意为骨髓，而不是西葫芦）。此外还有一种使用番红花给苹果调味的食谱：

苹果派和苹果卷饼

制作苹果挞

准备上好的苹果和香料，还有无花果、葡萄干、梨酒和葡萄酒，充分捣碎之后使用藏红花上色，然后装进油酥面皮，放入烤箱进行烘烤。

还有一种历史悠久的水果派值得特别注意一下，这种派使用的是价格不菲的进口外来水果——橙子，且橙子要用昂贵的糖来腌渍。橙皮蜜饯自伊丽莎白时代到18世纪一直非常受欢迎。橙皮蜜饯派是一道价格不菲的美味，有时会添加苹果分层。伊丽莎白一世的两个糕点主厨曾在1600年各做了一个橙皮蜜饯派，十分自豪地将其作为新年礼物献给了女王。由此可见，这种派是一种与女王身份相匹配的派。

我们在保存派的种类这件事上太疏忽了。那些添加无花果和藏红花的梨苹果派以及橙皮蜜饯派的消失对我们来说是巨大的损失。我们又用什么取代了它们呢？现如今的甜点派已经完全超出了水果和蛋奶酱的范畴，派和蛋糕之间的界限也变得模糊，有些派就像是带皮的蛋糕（脑海中马上浮现了山核桃派），甚至还有一些甜派是用蔬菜制成的。

蔬菜派

历史上的烹饪书籍中也出现过蔬菜派。不过几个世纪以来，蔬菜在人类饮食中的角色和重要性已经发生了巨大变化。除了恪守教规的某些宗教团体成员之外，我们那些中世纪的祖先应该很难理解，居然有人心甘情愿地放着肉不吃，去吃蔬菜。在中世纪时期，尤其是在冬天的欧洲，高蛋白食物一直都是非常稀缺的资源，人们从不嫌多。

一些古老的烹饪书里提到过很不错的蔬菜派，不过这些蔬菜派和当时的水果派一样，并不一定完全是 vegetarian（素食），vegetarian 是一个相对比较现代的词。现在我们做蔬菜派和水果派的时候可能会用到黄油，而那时使用的常常是骨髓，比如下面这个来自 1720 年左右的食谱：

洋蓟派[①]

取 6 个或者 8 个洋蓟，煮熟之后切片，使用甜味香料进行调味，再与 3 根骨头的骨髓和香橼混合。添加柠檬皮、橘黄色植物根茎、西洋李子、醋栗。随后再添加一层葡萄、香橼、柠檬以及黄油，最后把派封起来。胡萝卜派和土豆派的做法相同。

蔬菜派并不一定是咸味的。传统的感恩节南瓜派具有悠久的历史。17 世纪就有些食谱提到过"南瓜派"〔pompion pie，pompion 是英语中"南瓜"（pumpkin）的旧名〕，除此之外，诸如甘薯、欧亚泽芹（胡萝卜家族的一员）等甜味蔬菜同样长期以来被用作蔬菜派的常用馅料食材。另外，

---

① 洋蓟，也称朝鲜蓟、菜蓟。——编注

蔬菜派还会把甜味和咸味结合起来。1750年，威廉·埃利斯的农业和家务手册中有一份"洋葱派"食谱。这种派的馅料一半使用洋葱，一半使用苹果。不过，食谱并没有说明哪样食材是另一食材短缺时的替代品，也没有提到这种派应该什么时候吃。

那些并非素食主义或纯素食主义（veganism，一个更新出现的词）的群体可能会被迫食用素食，比如大多数社会中的农民、大斋节期间的忠实信徒以及战时的普通民众。在缺少食材或者某些食材遭到禁食的情况下，厨师一般有两种基本选择：一是彻底接受替代食材，发挥创造力；二是努力让替代食材在外观和口味方面尽可能接近原版食材（可能会告知食客，也可能不会告知）。在二战期间，英国不得

南瓜派

不依靠小麦支撑经济[1]，并且对肉类进行了定量配给。这样一来，工人们晚餐的重头戏——派，将何去何从？第一代伍尔顿伯爵弗雷德里克·马奎斯曾于1940年出任粮食大臣，负责组织食物定量配给工作。马奎斯将这项工作完成得十分出色，实现了一个看似不可能完成的壮举，同时赢得了英国民众的大力拥戴。在没肉吃的日子里，有一种蔬菜派的食谱就是以他的名字命名的。根据描述，这种派"看起来和吃起来都很像牛肉腰子派，只不过没有用到牛肉和腰子"。这次名人背书更像是一种宣传推广上的成功，而非烹饪领域的成就。究竟如何，请你自行判断。

伍尔顿派

　　取土豆、花椰菜、大头菜、胡萝卜各一磅，切碎，外加三四棵大葱。条件允许的话，可以添加一茶匙蔬菜提取物和一汤匙燕麦。把所有食材放在一起煮，水量没过食材即可，偶尔搅拌一下，避免粘锅。食材煮熟冷却之后放进派烤盘里，撒上切碎的欧芹，最后覆盖一层土豆外皮或者全麦油酥面皮。把烤盘放进火力中等的烤箱中烘烤至油酥面皮呈美妙的棕色，配上棕色肉汁趁热食用。

---

[1] 二战前英国用的小麦大多是从加拿大进口的加拿大小麦，但因运输这种小麦会占据原可以装载军火等物资的货舱空间，所以英国开始大量种植口感较差的英国小麦以自给自足。——编注

不过，蔬菜派并不需要成为低肉派一头的替代品。略带神秘色彩的法国男爵布里斯曾经在他 1868 年的烹饪书里展示了一份蔬菜派食谱，这种蔬菜派采用了开放式水果挞的形态，使用油酥面皮条进行了区域分割，每个区域都填满了不同颜色的蔬菜。正如男爵指出的那样，这种派"不仅是味觉上的盛宴，还是视觉上的享受，能很好地为我们斋戒日里的餐桌增色添彩"。

简朴的派

几个世纪以来，家庭生活的节俭艺术一直备受烹饪书籍作家的赞誉。美国作家莉迪亚·玛利亚·蔡尔德在她的作品《节俭的家庭主妇：致不以节俭为耻的人们》（1830 年）中对此进行了很好的总结：

> 在操持家庭方面，真正的"节俭"其实就是收集所有碎片，这样可以避免各种损失……任何东西只要还有利用价值，就不应该丢掉，无论这种价值多么微不足道。

就本质特点而言，派非常适合用来践行这种艺术，不过有时可能过犹不及。查尔斯·狄更斯曾经在《我们共同的朋友》中生动地描述道：

晚餐过后，盘子里的残羹剩饭连带剩下的凝固肉汁被一起倒回了剩下的派里，日常节俭中的一项经济投资就完成了。莱德胡德给自己倒了一大杯啤酒，痛快地喝了起来。

狄更斯的同时代美国作家纳撒尼尔·霍桑对于派——至少对于英国旅馆提供的派——同样不太有把握：在英国的旅馆，"有时你吃下一个肉派之后，可能会有沉重的心理负担，感觉自己吃了别人晚餐的残羹冷炙"。

这种将一周的剩饭剩菜装进派的做法一度极为常见，这些派甚至还拥有了特定的名称——"剩饭派""周六派""老姑娘派"，对于这些派的馅料，劝你不要抱有任何幻想。

节俭（不包括吝啬在内）可以说是一种家庭美德，但

端着圣诞派

二战时期的马麦酱广告

是对于很多人来说却一直是一种过日子的必要手段。对于那时最为贫穷的人群来说，最简朴的派莫过于"刚毛派"。根据18世纪早期一份"为了改善畜牧业和贸易现状"的手册的描述，人们屠宰牲畜并把皮毛运送到市场。在市场里，"贫穷的妇女……还会切下牛角旁的细碎牛肉，用这种叫作'刚毛'的碎肉来制作派"。

某些时期还出现过必须在全国范围内践行节俭生活方式的情况。二战期间，美国曾掀起一场节俭运动，要求把小麦消耗量减少40%，脂肪消耗量减少20%；对美国家庭主妇们的要求之一就是制作"无覆盖"的派。18世纪，英国遭遇了小麦的严重短缺，而这也产生了一些有趣的长期影响。正如我们已经知道的，当时派在餐桌上的"存在"非常重要。为了弥补小麦短缺时派的消失，当时那些了不起的陶器制造商模仿派的构造设计了一种边缘呈扇形且上了金釉的盘子，也就是所谓的"派皮陶器"。这种"陶器派"就是早期版本的锅派。小麦短缺所带来的另一个积极影响在于，不加外皮的派馅料逐渐走上了独立发展之路，演变成了肉糜、肉泥和肉酱。

骗人的派和险恶的派

派的馅料隐藏在外皮之下，因此派自带神秘感。这种神秘感可以是成功之兆，也可以是一种诅咒。通常我们想要的是惊喜，而不是惊吓；关键的问题在于，切开面前这个

《包厘街的狗派》：这幅漫画利用了人们对派馅料与广告宣传不符的普遍恐惧，且带有种族歧视色彩。

《伦巴底街的事故》（1787年，查尔斯·威尔逊·皮尔绘）

派的时候，我们是打开了一座神秘宝库，还是一个潘多拉魔盒。这里最让人担忧的就是肉派。勤俭节约、无害欺骗和险恶用心之间的界限会在肉派的制作中被模糊，好在我们还有法律的保护。

澳大利亚法律规定，肉派必须至少达到 25% 的含肉量。这个数量听起来很少，不过当你了解到这些肉派的馅料里可能包含了牲畜鼻子、耳朵、蹄筋、血块、血管以及很多其他部位的时候（不会包含动物胚胎和内脏，这一点可以放心），你会感到更加忧虑。事实上，肉类的检测方法其实是在检测蛋白质，因此使用动物血块制成的派完全有可能通过含肉量测试。

在家庭场景下，当主人把一种食物当作另一种食物来招待客人，并且可能是有意欺骗时，这背后藏着一系列不同的动机。这样做大概是为了给客人留下好印象，这种情况下的欺骗最好是好心的，且别被客人发觉。1660 年 1 月 6 日，塞缪尔·佩皮斯在堂兄弟家吃了一顿饭，他显然对这顿饭没有什么好印象，因为"鹿肉馅饼里用的明显是牛肉，一点也不大方"。还有一次，他吃到了一个"被证实是咸猪肉馅饼"的鹿肉馅饼。17、18 世纪的烹饪书籍里经常会出现使用牛肉仿制鹿肉的食谱，所以这一类现象在当时可能很普遍。

水果派同样也会出现造假现象。比如在澳大利亚，人们的确偶尔会听说"苹果派"里用的完全是佛手瓜，但这种情况至少仍然使用了植物作替代品。20 世纪 30 年代的乐之饼干生产商在产品销售方面十分大胆，几乎没有人可以与之相匹敌。多年以来，他们一直冠冕堂皇地在包装上展示"仿苹果派"食谱，这份食谱的馅料完全是用饼干制成的。令人难以置信的是，这种派在当时居然广受欢迎，几乎成了人们狂热追捧的时尚。更令人难以置信的是，很多狂热爱好者都声称它与真正的苹果派别无二致。

与那些使用病畜肉和假肉做成的肉派相比，假水果派显得不值一提。那些使用牲畜鼻子或者尾巴做成的肉派，或者艾伯特·史密斯在《伦敦生活与特质速写》（1849 年）里提到的"包裹之下有着猫肉特征的不明之物"还不是最过分的。有一种更糟糕的肉叫作"机械再生肉"，这是一种从死畜身上刮下来的肉，用来制作汉堡和派。这种泛红的肉糊疑似会传播疯牛病——没有比这更险恶的馅料了，不是吗？

有一种原始恐惧为很多文学作品和都市传说提供了素材——用人肉做成的派。莎士比亚在他的极度血腥之作《泰特斯·安德洛尼克斯》（或称《圣诗复仇》）中，把这种骇人的想法用作了复仇的终极武器。作品中的泰特斯是一名罗马将军，他杀死了宿敌哥特女王塔莫拉剩下的两个儿子，用他们的肉做成"两个巨大的派"，让女王在不知情的情况下吃了下去。在《理发师陶德》这个故事中，一名伦敦理发师通过割喉的方式杀死了自己的很多顾客，然后让情妇把这些尸体做成肉派卖给了食客。这部作品为众多读者与观众带来了长达一个世纪左右的恐惧和惊骇，而且该作品的原型有可能是 1785 年伦敦当地的一宗谋杀案。

历史上充斥着饥荒和极度艰苦时期人吃人的故事。这一类故事让我们感到不适，但与此同时我们也会为之着迷。人肉到底是什么味道呢？有传言说人肉尝起来很像猪肉。也许正因如此，17 世纪与 18 世纪早期盛极一时的"美人鱼派"使用了猪肉做馅料。在当时，美人鱼是一种令人极为着迷的存在，人们对所谓的目击事件产生了极大的科学兴趣。一些神职人员针对食用美人鱼肉是否可以算作吃人肉展开了激烈的辩论，因为美人鱼毕竟具有人类的一半特征。在那一个多世纪中，很多烹饪书籍都提到过美人鱼派的食谱，制作方式基本没有发生过变化。以下为威廉·萨蒙的作品《家庭词典，或家居伴侣》（1695 年）中的版本：

美人鱼派

准备一头猪，烫洗、剔骨。用布把肉擦干之后，使用碾碎的肉豆蔻、胡椒以及鼠尾草末进行调味。准备两条牛舌，煮熟后晾干、冷却，切成半克朗硬币厚度的长条肉片。把四分之一的猪肉铺进方形或圆形的派里，再铺上切片的牛舌，如此重复四次，最后在顶部铺上培根肉片，撒些丁香，放入几片黄油和月桂叶之后即可进行烘烤。烤制过后用甜黄油填满派，然后使用黄油和花朵把派装饰成白色，最终就制成了猪肉派，即所谓的美人鱼派。放冷之后食用。

从另一方面看，有一种与苏格兰关系十分密切的派叫"油炸派"，外观令人不敢恭维且对身体健康有害，因此同样可以算作险恶的派。这种派在烤熟之后还要进行二次油炸。由此可见，苏格兰的"勇士之地"称号并非无中生有。

用心险恶的人肉派，由伦敦舰队大街的恶魔理发师提供人肉馅料。图片为电影《理发师陶德》（2007年）剧照。

一名烘焙师端着一盘派。图为根据威廉·荷加斯的作品《向芬奇利的进军》（1750年）制作的雕版画。

特殊场合中的派

> 在我看来，如果没有派和烤肉的话，一场宴会
> 就没有办好。
>
> ——约翰·泰勒（1578—1653）

派具有很强的实用性，功能多样，广受喜爱，而且还自带一个易于装点的可食用的油酥面皮礼盒，也难怪派仍然在许多我们最喜爱的庆典之中扮演着重要角色，甚至成了很多活动的象征。

圣诞派

最初有一种牛奶小麦粥，这种朴素的粥是农民的主食，富人则把这种粥当作搭配鹿肉的配菜。在圣诞节之类的特殊场合，人们就会使用糖和香料给小麦粥丰富口味，还会添加许许多多其他的好东西，比如鸡蛋、果干（梅子）、葡萄酒和肉末。这种圣诞粥（或浓汤）在放入油酥面皮盒烹制的情况下，最终演变为圣诞（百果）派①；在用布包裹起来烹制的情况下，成为圣诞布丁；使用特定形状的金属罐进行烹制的，就成为圣诞蛋糕。

虽然这种百果派可能已经不再使用肉馅，且其他原料可能现在一年四季均可获取，但它却并未因此而失去与圣

---

① 百果派的英文为 mincemeat pie 或 mince pie。mincemeat 由 minced meat（肉末）演变而来，译为百果馅，由果干碎、烈酒、香料、牛板油或植物起酥油混合而成。过去人们也会在百果馅中加入牛肉或鹿肉。——编注

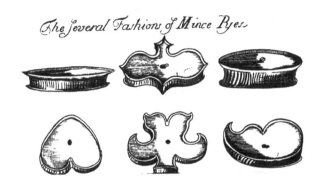

The Several Fashions of Mince Pyes

此图出自《英格兰最新烹饪方式与油酥面皮制作方法》（1708 年），亨利·霍华德著。

诞节的联系。17 世纪的清教徒把这种派称为"包裹着外皮的偶像崇拜"，一度极力试图禁止它，但没有成功——圣诞百果派留存了下来。而另一类圣诞派就没有这么好运了，这类派尺寸很大，外观为盒状，使用上好的肉制成，非常结实。其中最著名的一种来自约克郡，现存最早的食谱来自汉娜·格拉斯的《烹饪艺术》：

如何制作约克郡圣诞节派

首先做一个质量较好的立式油酥面皮外壳，壁和底要做得很厚。准备一只火鸡、一只鹅、一只家鸡、一只鹧鸪、一只鸽子，剔骨之后进行调味。取半盎司[1]肉豆蔻衣、半盎司肉豆蔻、四分之一盎司

---

[1]　1 盎司约等于 28.35 克。——译注

丁香、半盎司黑胡椒，碾碎混合在一起。然后取两大勺盐，与调味料搅拌在一起。给禽类剔骨的时候沿背部切开，先给鸽子剔，然后给鹧鸪剔，用东西盖住剔好的肉；之后再按照家鸡、鹅、火鸡的顺序剔骨，且选取的火鸡一定要够大。给所有肉调味之后把它们按同样的顺序放入油酥面皮，让馅料整体看上去像一整只火鸡。准备一只野兔，用干净的布擦干净，切成大肉块，调之后紧密摆放在馅料两侧。接下来准备山鹬和其他野味，以及你能搞到手的野禽，调味之后紧密摆放。往派里添加至少四磅黄油，然后盖上派皮。这里同样需要很厚的派皮。最后把整个派放进温度很高的烤炉里，至少烘烤四个小时。

制作油酥面皮大约需要一蒲式耳①面粉。本章节后面将介绍油酥面皮的制作方法。人们通常会把这些派装进盒子里，作为节日礼品寄往伦敦，因此派壁必须坚固。

感恩节

在美国，人们已经毫不含糊地发展出了一种食用甜派的饮食传统，这一点在每年 11 月份的感恩节展现得尤为淋

---

① 蒲式耳，固体计量单位，1 蒲式耳约等于 36.37 升。——译注

漓尽致。在感恩节期间，整个美国都在为派而疯狂，而传统的感恩节派也表明，这是丰收的季节。当然，地域差异是不可避免的。20世纪50年代，美食作家克莱芒蒂娜·帕德福德声称自己总结了各地之间的差异："告诉我你祖母来自哪里，我就能说出你感恩节会吃多少种派。"帕德福德女士说，来自美国中西部地区的人会吃两种派——百果派和南瓜派；来自新英格兰地区的人会吃百果派、南瓜派、蔓越莓派这三种派；波士顿人则吃四种——百果派、南瓜派、蔓越莓派和苹果派。帕德福德女士没有提及美国西部地区，但也没有解释具体原因。而美国南部地区则因为一些比较

感恩节派

有争议的缘由，不喜欢派。

任何有关感恩节的讨论都没有提到咸味的派。美国人对于甜点派有着绝对的迷恋，这一点非常有趣，我们将在第六章进一步探讨。

乡村派

英国每一个村庄似乎都有一个特定的日子来庆祝当地的传统，而庆祝时的食物经常都是派。几个突出的例子就足以说明这一点。出于已被人遗忘的古老原因，在复活节后的星期一这一天，英国莱斯特郡哈勒顿村当地的牧师必须在早礼拜之后向信众发放一个野兔肉派。剩下的派会被拿到村庄郊外一个叫"野兔肉派滩"的地方扔向人群，随后人们便开始玩起狂野的传统游戏——踢瓶子。

康沃尔郡有一个叫作"毛斯尔"①的渔村，当地保留着一种在 12 月 23 日晚上吃"仰望星空派"的传统习俗。这种派使用了沙丁鱼，鱼头从派中央的油酥面皮中探出，似乎在仰望星空，十分有趣。这种派是为了纪念当地的传奇英雄汤姆·鲍科克（Tom Bawcock，bawcock 是个旧词，意思是"好人"）。据说在一个气候恶劣的时节，汤姆·鲍科克曾在一个天气恶劣的晚上外出并带回了足够的鱼，使村民们免于饥饿。

---

① 英文名为 Mousehole，字面意思为"老鼠洞"。——译注

来自康沃尔郡的传统仰望星空派，鱼头露出油酥面皮表面。

在英国德文郡的拉普福德堂区，当地人曾经在 7 月 7 日纪念圣托马斯·贝克特的节日大吃"杵派"（pestle pies）。这里的 pestle（杵）指的是动物的大腿，杵派则是一种尺寸较大的立式派，馅料"包含一整条腌猪后腿，有时也会有一条牛舌，还会包含两三只家禽，或者一只火鸡也行"。12 世纪，当地一名参与了 1170 年大主教谋杀案的贵族为这位圣人建造了一座教堂来赎罪[1]，但至于圣托马斯·贝克特与猪腿派之间的关联，目前尚不清楚。

---

[1] 圣托马斯·贝克特曾担任坎特伯雷总教区总主教，但因反对亨利二世对教会的干涉而触怒了亨利二世，于 1170 年被后者支持的四位男爵骑士在坎特伯雷大教堂刺杀而殉道。教皇亚历山大三世于 1173 年封其为圣人。——编注

用来履行义务的派

在过去，人们经常会把食物作为象征性的礼物或者象征性的付款方式。派可以做得引人瞩目、精美奢华，而且容易保存和运输，因此很适合这些用途。

依据古老的习俗，格洛斯特市曾在圣诞节期间向君主献上七鳃鳗派以表忠诚。七鳃鳗是一种类似鳗鱼的无鳞淡水吸盘鱼，因肉质肥美而在过去备受青睐。向王室赠送七鳃鳗派的传统终结于维多利亚女王统治末期。不过在1953年女王伊丽莎白二世的加冕礼上，英国皇家空军餐饮部队制作了一个重达42磅的七鳃鳗派，重现了这一传统。

数字"24"经常会出现在派的历史中。民俗学家们对于童谣"二十四只黑画眉被烤成了派"的具体含义仍然争论不休。"24"相当于"两打"，"两打"在许多文化中都是一个意味深长的数字，尤其是在具有象征意义的送礼场合，送出的鲱鱼派数量也是"两打"。这种现象背后的起因很难解释。

不过，选择鲱鱼派作为礼物的原因显而易见：鲱鱼派油多肥美，易于保存，而且很适合斋戒日食用。雅茅斯镇因鲱鱼而闻名，根据古代宪章的规定，当地曾经每年需要为国王献上24个鲱鱼派。诺维奇市的治安官同样有义务在大斋节向当地的凯斯特领主献上24个鲱鱼派。

最后一个例子是关于个人义务的。爱德华兹家族是一个牧羊人家族，这个家族在康沃尔郡哈利恩村圣康斯坦丁

唱一支六便士之歌，一口袋黑麦

二十四只黑画眉被烤成了派

打开了派，鸟儿开始歌唱

把这美味佳肴献给国王[①]

教堂教区的一个小屋里生活了几百年。房屋租金只要每年一个帽贝派（内含葡萄干和香草），在 3 月 9 日的圣人节缴纳。帽贝曾经是穷人的食物，因为在海边的潮汐线上就可以捡到。由此可见这笔租金确实非常低廉。

新娘派

在结婚蛋糕出现之前，曾经有过"新娘派"。结婚蛋糕并

___

① 英国童谣《六便士之歌》。——译注

非由新娘派演变而来，而是与圣诞蛋糕有着相同的起源。新娘派则是另外一种完全不同的东西，最初的新娘派并没有特定食谱或者强制使用的原料，只要使用当时能获取的顶级美味食材制作派的馅料即可。这种派常常被称为"巴塔利亚派"（batalia pie），这个名字来源于法语 beatilles[1]，意思是"漂亮的小东西"，比如鸡冠、小羊睾丸、鹅内脏等等。厨师们显然不了解这个名字的来源，误以为这个名字与城垛（battlements）有关，所以经常会把巴塔利亚派做成城堡造型，还要搭配上塔楼。无论有没有塔楼，巴塔利亚派都是一种完全适用于庆典的食物。一些早期的巴塔利亚派食谱还明确指出，这种派非常适合用作新娘派。

最令人惊艳的新娘派食谱来自罗伯特·梅的烹饪书《卓越厨师》。17 世纪是英国派的鼎盛时期，而这种新娘派绝对称得上这个时期的巅峰之作。

如何做出底部由不同派组成的
非凡的派（新娘派）

准备雄鸡睾丸和鸡冠，或小羊睾丸和小牛胰脏，用热水焯一下，切块。准备两三个公牛蹄，热水焯过之后切片。另外准备一品脱牡蛎、切片椰枣、一把松子、少许腌制过的金雀花芽、一些肥瘦相间的培根片、九个或十个烤过且用热水焯过的栗子。使

---

[1] 原文有误，法语为 béatilles。——编注

用盐、肉豆蔻、大块的肉豆蔻衣给所有食材调味，然后用一点黄油把派封起来。准备一些黄油、两三个蛋黄、一些白葡萄酒或者红葡萄酒，加上一两个柠檬的汁水搅拌成酒汤。切开顶面油酥面皮，倒入汤汁，摇匀。在馅料上摆放柠檬片和腌制过的浆果，再把馅料重新包裹起来，放进派的中央或扇形饰边。

再做几个与第一个形式相同的派，不过这里需要确保三种样式互相成比例。可以把几个派放在同一块油酥面皮底上（这种方式操作起来比较便捷），也可以把它们分开。在分开的情况下可以烘烤一下中间那块面粉，烘烤完毕凉下来之后，掏空底部面粉，放入几只活鸟或者一条蛇，这样客人切开派的时候会感到很奇特。这种做法只是为了在婚礼上消磨时间。

另外几个派可以使用若干种食材制作馅料。比如做其中一个派，可以准备一些有须的牡蛎，煮至半熟后用大块肉豆蔻衣、胡椒、少量碎姜和盐进行调味。把牡蛎填入油酥面皮之后再抹上骨髓和一些质量较好的黄油，封起来放进烤炉烘烤。制作调味汤汁要用到白葡萄酒、牡蛎汁及一整个洋葱，或者事先用一瓣蒜涂抹容器底部；把三四个牡蛎切碎放进去可以使汤汁更浓，不过煮完要把牡蛎肉捞出来。煮的时候加入一块黄油和一个柠檬，加一捆扎紧的香草一起煮也很不错。把派的顶部油酥面皮切开或者开一个孔，倒入汤汁。

第二个派可以用虾和鸟贝做。先给虾和贝肉进行调味，不放骨髓。手头有腌蘑菇的话可以烤几个，外加一块搅打过的黄油、少许醋、一颗切成片的肉豆蔻和两三个橙子的浓郁汁水制作汤汁。把汤汁倒进派里。

第三个派可以做鸟肉派。准备几只雏鸟（例如云雀雏鸟），拔掉羽毛，去掉内脏。使用面包碎、香草细末和牛油（或者切碎的骨髓）制作五香碎馅，把杏仁混合些许奶油碾碎可以防止出油，还可以加入一点帕尔玛意大利干酪或者陈年奶酪（也可以不加）。使用肉豆蔻、姜、盐给碎馅调味，然后像制作布丁那样加入奶油和鸡蛋一起搅拌。把馅填进云雀肚子里，然后使用肉豆蔻、胡椒、盐给云雀调味，摆进派里。派里放一些黄油，在云雀之间撒一些松子、蛋黄以及香草，蛋黄和香草需要切得非常碎。把派放进烤炉里烘烤，与此同时使用橙汁和搅打过的浓稠黄油制作调味汁，摇晃均匀。

至于另一个派，可以把洋蓟煮熟，取下菜心，把每个菜心切成不多于四块的大小，再使用肉豆蔻调味。这样一来我们就可以使用好几种原料来制作其他几个派的馅料。

婚礼之后便是家庭的诞生。在过去，人们认为很多家庭活动都需要做一个特殊的派或者蛋糕来庆祝。过去有一种派叫作"呻吟派"，用来在"呻吟时间"招待参与者和客人。"呻吟时间"指的是分娩，这种称呼虽然古旧，但也十分贴切。我们现在已不再制作"呻吟派"和"洗礼派"，不过宾夕法尼亚州的荷兰裔仍然保留着"葬礼派"。葬礼派使用的主要原料是葡萄干，因此也被称为"葡萄干派"或者"罗西娜派"①。原料选用葡萄干可能是因为葡萄干无论在什么时候都很容易获取。这种派也很方便运输。

巨型派

烹饪的历史中有很多超大型派的故事，其中很多主角是圣诞派。你可能会觉得格拉斯夫人的派已经很大了，但英国谢菲尔德市的酒馆老板罗伯茨先生在1826年做的一个派会让你更加吃惊。这个派"使用了大量兔肉、小牛肉以及猪肉，烘烤之前的重量高达15英石10磅②"。1835年，英国罗瑟勒姆镇旧船旅馆的柯克夫人做了一个更大的派，"烘烤之前重量高达17英石，使用了一块牛臀肉、两条小牛腿、两条猪腿、三只野兔、三对家兔、三只鹅、两对野鸡、四对鹧鸪、两只火鸡、两对飞禽和7.5英石最优质的面粉"。

英国约克郡有个小村庄叫作登比代尔，村民或多或少都对巨型派有所了解——当地的巨型派制作历史已经超过

---

① "罗西娜"（rosina）为德语中"葡萄干"一词的音译。——编注
② 1英石等于14磅，15英石10磅约等于99千克。——译注

了 200 年。第一个巨型派出现在 1788 年，是为了庆祝英国国王乔治三世从一次疯癫中康复而制作；第二个制作于 1815 年，是为了庆祝滑铁卢战役取得胜利；第三个制作于 1846 年，为了庆祝《谷物法》①的废除。还有两个巨型派制作于 1877 年，是为了庆祝维多利亚女王登基 40 周年。这两个派中的第一个由专业烘焙师制作，重量达 1.5 吨，不过后来由于肉馅变质无法食用，很快就被埋到地下；一个月之后，当地的一群家庭主妇又做了一个"复活派"。1928 年，当地人做了一个派为医院集资，这个派长 16 英尺，宽 5 英尺，高 15 英寸，油酥面皮使用了 1120 磅面粉和 200 磅猪油，馅料使用了 1500 磅土豆和 6 头阉小公牛的肉。1964 年，当地为了集资建造村礼堂，又做了一个派。"登比代尔 200 周年纪念派"制作于 1988 年，长 20 英尺，宽 7 英尺，高 18 英寸。最近的一个巨型派出现在 2000 年，具有三重庆祝意义：这个 12 吨重的"千禧年派"也是为了纪念伊丽莎白王太后②100 周年诞辰，以及当地铁路开通 150 周年。

具有娱乐功能的派

可能有人会认为，所有的派都具有一定的娱乐元素，因为每个派都是用外皮包裹起来的一份惊喜。中世纪的"城

---

① 《谷物法》是 1815 年至 1846 年英国强制收取谷物进口关税的一部法案，目的是维护土地贵族的利益。——译注
② 前任英国女王伊丽莎白二世的母亲。——译注

堡派"——城堡造型的派（详见第36页），想必在有趣的同时也十分美味。不过还有一些派属于一个完全不同的范畴，它们的功能显然在于娱乐，也显然不是用来吃的。这种派的成功在于为每一位食客带来预期的欢乐。下面这份来自16世纪意大利的食谱就是一个例证：

如何制作容纳活鸟、切开之后鸟儿飞出的派

做出一个大号的油酥面皮盒，在底部掏一个不小于拳头大小的洞，油酥面皮盒高度需要比普通的派高一些。把油酥面皮填满面粉放进烤炉烘烤，烘烤完再打开底部的洞，取出面粉。根据油酥面皮盒底部空洞的大小制作一个派放入油酥面皮盒，然后还要在这个派的周围放上活鸟，油酥面皮盒剩余的空间能塞多少只就塞多少只。准备就绪之后把整个派端到客人面前。揭开或者切开这个巨型派的顶面油酥面皮时，所有的鸟就会飞出来，为客人带来乐趣和愉悦。由于不能完全欺骗客人，此时应该切开油酥面皮盒里的那个小派。小派也可以换成许多其他的糕点，比如水果挞。

让鸟儿从派里飞出来取悦客人的做法曾流行了几个世纪。《卓越厨师》的作者罗伯特·梅则更胜一筹，他曾解释过一种无比复杂的派（油酥面皮）结构的做法，这种派可以容纳青蛙，青蛙跳出的时候，"在场的女士会尖叫着跳

起来"。

在一个人们完全不关注政治正确性的时代，这种创意有过几次进一步的飞跃。杰弗里·哈德森[①]生于 1619 年，9 岁那年身高只有 18 英寸（尽管"身材比例匀称"）。他之后的人生经历刺激无比：参与过决斗，被海盗俘虏，还曾作为政治犯锒铛入狱。但他最初名噪一时是因为曾经作为惊喜礼物藏在派里。英王查理一世和王后亨利埃塔·玛丽亚就是接受这份惊喜礼物的幸运儿。他们被逗得大乐，随即就"收养"了哈德森（即收他为宠儿）。

1895 年，美国纽约出现了对这一主题的另一次诠释。当时，纽约当地的一群上流社会男士受邀参加了一次秘密晚宴，据说晚宴上还会有一道神秘的佳肴。结果这道佳肴其实是一个派，派里有一大群金丝雀，紧接着还出现了一个名叫苏茜·约翰逊的 16 岁女孩，身着毫无遮掩性可言的衣物。这件事的后续从未得到证实，但毫无疑问，（没被邀请的）公众对于"女孩派晚宴"的整个概念感到非常愤怒。

① 杰弗里·哈德森（Jeffrey Hudson，1619—约 1682），亨利埃塔·玛丽亚王后的宫廷侏儒，曾被称作"王后的侏儒"以及"矮人勋爵"。1626 年，小哈德森被献给了白金汉公爵夫人。几个月之后，公爵与公爵夫人在伦敦宴请英王查理一世和亨利埃塔·玛丽亚王后。在这次宴会的高潮部分，王后面前摆上了一个巨大的派，身高 46 厘米的哈德森身穿小型盔甲突然从派里站了起来。王后非常开心，于是公爵和公爵夫人把哈德森作为礼物送给了王后。——译注

［第六章］

和派一起环游世界

不要轻蔑地把菜肴仅仅当作食物，这种神眷之物本身就是整个文明。

——阿布杜勒哈克·西纳西

我们人类一直在不断地向世界各地迁移，并且在迁移过程中保留着自身的饮食习惯。人类这样做是为了利用自己的农业与烹饪知识，并且食用熟悉的食物可以维持我们自身与故乡的联结，缓解乡愁。有时候我们可能不得不更换食材或者调整烹饪方法，但即使经过了几代人，我们继承的传统仍然会在自家的菜肴里明显体现出来。

当欧洲人开始从他们原本的定居地迁往世界各地之时，他们也向各地带去了以谷物为主要原料的美食，尝试制作他们熟悉的面包和糕点。17 世纪的英国在海事力量与烹饪技术两方面都达到了鼎盛。随着英格兰人帝国的扩张，派也随之传播，并且根据不同地区的当地食材和环境条件进行了改良。

热衷于派的英国人曾掀起两次移民潮，一次是从 17 世纪早期开始迁往美洲，一次是在 18 世纪迁往澳大利亚；由此便又诞生了两个热衷于派的国家。有趣的是，虽然同样起源于英国，派在美国和澳大利亚这两个前殖民地国家所代表的概念却截然不同。在美国，单单说"派"这个词时，毫无疑问就是指一种甜点；而澳大利亚的"派"指的就是肉派。

这一差异在这两个国家举办的做派比赛中得到了淋漓

尽致的体现。美国派协会举办的"全国年度派锦标赛"共包含12至14种项目分类，全部都是甜口。苹果、水果（或浆果）、奶油、柑橘、蛋奶酱、南瓜、甘薯、坚果、巧克力酱以及花生酱都是比赛固定的口味类型，此外还有一些其他类型。相比较而言，2006年举办的"澳洲派大赛"（比赛仅面向商业烘焙师）只包含了5种项目分类——红肉、禽肉、野味、鱼肉以及素食，感受不到一丝一毫糖的气息。

由此可见，同一烹饪传统在两国的发展道路并不相同。为什么会出现这种现象呢？

吃派比赛（1945年7月4日，美国）

美 国

　　第一批大规模抵达北美的移民是一群清教徒[①]，他们希望成为小型农户，并带来了所需的生产设备，而事实证明，他们选对了定居地点。特别是那些在"老英格兰"深受人们喜爱的苹果在新英格兰地区生长得非常好，几十年之内就成为当地的主要产品之一。脱水之后的苹果干也成了当地人在前往西部地区遥远路途中的标配食物，而苹果干的主要用途就是制作苹果派。

　　在移民早期阶段，小麦的收成情况远远不及苹果。他

威廉·亨利·谢尔顿的一幅油画作品（约 1863 年），描绘了南部联盟士兵在一次突袭中从敌对阵营的农舍里拿走自制派的场景。

[①]　原文为 Puritans and Pilgrims，两者都起源于英国国教，都是清教主义团体。——编注

具有国家象征意义的派：这幅漫画描绘了一个英国人（可能是贪吃的乔治三世）在狼吞虎咽地吃一份象征法国的派，代表着他要击垮拿破仑·波拿巴。图片出自乔治·伍德沃德于1803年创作的蚀刻版画。

们后来才通过原住民了解到，这里是玉米的国度，不适合种小麦。直到西部地区被真正地完全征服之后，小麦才得到了大规模的种植。不过这个时候的美国人已经养成了节省小麦的习惯——手上只有少量小麦时，如果用它们去做派而不是面包，它们会更加耐用。

今天的新英格兰地区成为美国的派之都并不令人吃惊，早餐吃（甜）派在当地已经成为一种传统。其中苹果派更是深深扎根于美国历史——它与故国、新国和开拓精神相关联，永远与国家意识和爱国情感画上了等号。

澳大利亚和新西兰

第一批欧洲移民移居澳大利亚大多并非出于自愿。在当时，工业革命催生了大型城市，而这些移民来自城市中的贫民群体，由罪犯和海军陆战队员构成，大多缺乏务农和烹饪方面的技能。对于城市贫民来说，肉非常珍贵。如果他们手上有肉，那就很可能是从小饭馆或街边小贩那里买（或者偷）了肉派。这些最初的移民来到了一个非常适合种植小麦和放牧的国家，而后来的那批自由移民就是受到了"一天三顿肉"的诱惑来到了这片殖民地。

1972 年获选的悉尼市长曾经说过，澳大利亚这个国家建立在"肉派、香肠以及镀锌铁"之上。至少有一个事例可以证明这句话在字面意义上完全正确，而且相当具有讽刺意味。1927 年，堪培拉的新国会大厦举办开幕仪式，然而民众对于这整件事完全不感兴趣，因此餐饮承包商不得不把大量为活动准备的食物埋进了当地的垃圾场，其中包括 10000 个派。同年晚些时候，这处垃圾场之上建起了一座行政大楼，就是今天环境与水资源部的所在地——不知道为什么，这看起来似乎非常合理，尤其是当你得知大楼竣工后约有 620 吨用于地基的水泥并未被使用时。那些变质的肉派当时一定被认为足以构成大楼的地基了。

澳大利亚早期对于派的热爱似乎遍及各个社会阶层。1850 年，墨尔本《阿尔戈斯报》刊登的一篇文章提到了派，这篇文章也是澳大利亚本国报纸针对派的首次记载。文章指出，与会议厅里提供的食物相比，那些城镇里的市议员普遍更喜欢当地酒吧里的肉派。

除去非官方的认可之外，有一种独特的"漂浮派"已经被澳大利亚国家信托官方认证为"南澳大利亚遗产标志"。漂浮派是一种浸在豌豆糊里的肉派，外加番茄酱进行装点。大概很多人都会说，这确实是一个不错的遗产标志。特里·普拉切特的《最后的大陆》是一部隐约带有澳大利亚色彩的作品，他在作品中恰当地总结道："谁才是跨越这片红色沙漠的英雄？剪羊毛冠军，马背上的骑手，不停奔波的人，啤酒爱好者，丛林里的逃犯，还是那些在清醒状态下仍然可以吃一个漂浮肉派的人？"按照普拉切特的观点，漂浮派确实具有象征意义。澳大利亚人对派充满了真挚的敬意（与此同时，人们对其完全不抱幻想），这一点在一些关于派的俚语中得到了最充分的体现——他们把派称作"苍蝇公墓""老鼠棺材"，甚至"蛆袋"。

　　澳大利亚人自诩为世界上最大的派消费群体，不过这一殊荣很可能会落到新西兰人头上。大型的商业烘焙店对公开商业信息有所顾虑，而小型的家庭烘焙作坊又从来不去统计，所以精确的数据不得而知，永远无法正式确认哪

南澳大利亚著名的标志性美食漂浮派

个国家才是派的"最大消费国"。和在澳大利亚一样,在新西兰占统治地位的派也是肉派,而不是甜派;肉派是新西兰体育场馆里的传统快餐。新西兰非常适合养羊,因此羊肉派自然是这里最具有代表性的派。

加拿大

与其他的现代文化相比,加拿大的传统和文化或许更能从本国的派中清晰地展现出来。布雷顿角岛的"猪肉"派(见第47页)显然源于中世纪的欧洲传统,尽管现在可能已经不再包含猪肉。那么,这种派是19世纪上半叶逃离农业与社会剧变的大批苏格兰人带来的吗?还是那些可能为布雷顿角岛命了名的法国人带来的呢?法国的布列塔尼地区("布列塔尼人"与"布雷顿"的英文相同,都是Breton)以黄油和猪肉而著称,其他法国地区主要使用植物油。

加拿大的法国传统至少在两种魁北克派的名字上得到了体现:一种是叫作 tourtière 的双层皮肉派,在平安夜特别受欢迎;还有一种叫作 cipaille 或者 cipate 的派,包含多个分层,用肉、土豆以及油酥面皮制成。它的名称或许来自法语中的 six pâtés(六层派),而奇怪的是,这种派的构造和名称发音都与英国著名的"海派"(sea pie)极为相似。

最后,加拿大还有一种完全起源于本国的派,它因新斯科舍省的艰苦生存环境而诞生,叫作"海豹鳍派"。新斯科舍省早期专门捕猎海豹的渔民在捕获海豹之后，通常会把最好的部位卖出去，然后把别人不要的边角料留给家人吃，于是海豹鳍派就诞生了。据说吃这种派需要慢慢适应它的滋味。

其他地区

派有着无可争议的欧洲传统。不过在那些主要使用植物油的欧洲地区，烹饪风格必然会向其他方向发展。比如意大利就发明了披萨。pizza（披萨）这个单词常常被解释

《餐桌》（1925 年，皮埃尔·博纳尔绘）

为"派"，虽然可能不无道理，但我们必须明白，披萨使用的是面饼而非油酥面皮，因此披萨并不是派。如果把披萨算作派的话，那么烤三明治也可以称得上是一种派，那样未免太荒唐了。

人们一般不会把派（尤其是肉派）与经典法餐联系到一起。动物油和植物油法国人都非常善于使用，它们对于烹饪来说都非常重要。不过在他们制作经典酱汁的时候，黄油必不可少。中世纪时期，人们都是用外皮把食物包裹起来进行烹饪，而我们也已了解到法语中的 pâté（肉酱；肉派、馅饼）这个词与 pastry（油酥面皮、油酥糕点）之间存在关联，因此 pâté de foie gras（现指鹅肝酱）曾是一种填充了肥美鹅肝的派。但这并不意味着法国没有咸味的派——那里还有洛林咸味挞和阿尔萨斯山谷馅饼。不过法国的油酥面皮糕点一般都属于小型甜品类型，与蛋糕扮演着相同的角色。

德国人和奥地利人有着优良的烘焙传统，但似乎把精力都花费在了改进他们杰出的蛋糕工艺上。在那些土豆称霸咸味淀粉食物的国家，似乎都没有与派相关的重要传统。

继续向北，那里的派反映出了当地不利于种植小麦的寒冷气候和以鱼肉为主的饮食习惯。例如芬兰的卡累利阿派主要使用黑麦面粉制成，富含碳水化合物，馅料为大麦、荞麦、土豆或大米，搭配融化的黄油和煮熟的鸡蛋食用。芬兰还有一种"派"叫作"卡拉库克"，和披萨一样，它也不能算作真正的派，因为它是以黑麦面饼为外皮的"馅饼"，馅料混合了猪肉、鱼肉和培根。历史上曾经占领芬兰的俄罗斯有一种鸡肉派，是基督教节日的传统美食；还有一种更似饺子的三角小馅饼，炸透或者水煮之后食用。

还有一种类型的油酥面皮我们一直没有提及，但却不容忽视，那就是中东地区的著名美食——费罗酥皮（酥皮名的意思是"纤薄的叶片"）。人们会将它裹住坚果馅料、浸透蜂蜜糖浆，来制作一些让人欲罢不能的甜品，比如巴克拉瓦。不过费罗酥皮并不仅仅用于制作让人牙疼的甜点。有一种费罗酥皮派值得我们特别注意一下，那就是摩洛哥的巴斯蒂拉派。巴斯蒂拉派是一种甜味鸽子派，传承了早期的阿拉伯糕点制作工艺以及中世纪时期甜味与肉类相结合的饮食风格。

馅　饼

今天的我们会认为，馅饼是一种用手拿着吃的一人份半圆形咸味点心。然而，馅饼并非一直如此；在过去的年代里，"馅饼"常常指的是与今天完全不一样的东西。在过去的某些时期，派和馅饼之间并没有明确的区别，不过人们倾向于把那种包含了一整块肉（特别是鹿肉）的派称为"馅饼"，它们的尺寸通常很大。17 世纪中叶，塞缪尔·佩皮斯[1]曾经多次提到自己与几位朋友在几天时间内分食同一个

---

[1]　塞缪尔·佩皮斯（Samuel Pepys，1633—1703），英国作家和政治家，曾任英国皇家学会会长，是著名的《佩皮斯日记》的主人。——译注

鹿肉馅饼的经历。玛丽·蒂林哈斯特夫人在她的作品《稀有而优秀的食谱》（1690年）中说道："做牛肉馅饼的时候最好使用一块牛上腰肉；羊肉馅饼最好使用一块羊肩肉或者两块羊胸肉。一份鹿肉馅饼或牛肉馅饼需要烤制6个小时。"而这绝不是尺寸最大的馅饼——有些食谱还会要求使用1蒲式耳（36升）面粉，烹饪时间长达24小时。

今天的馅饼很适合工人群体，是可以拿在手里食用的完美餐食，非常便于携带到矿井或者田地里。它们与康沃尔郡之间存在着不可磨灭的联系，不过英国各地都能找到各式各样的馅饼，贝德福德郡馅饼就是一个非常好的例子。这种馅饼发源于其名字中的那个英国郡，一端填充着咸味馅料，另一端是甜味馅料，因此这种馅饼可以同时充当主菜和甜品两个角色。

康沃尔馅饼

康沃尔馅饼在古康沃尔语里叫作 oggie，其传统原料自然存在争议，不过大多数专家一致认为：馅料使用的肉为切碎的小肉块，而非绞碎的肉末；蔬菜（可能是土豆、洋葱和萝卜）必须切片；放入油酥面皮的原料必须是生馅料。卷曲的波浪状边缘是康沃尔馅饼的传统造型，拿着吃的时候这种边缘就成了"手柄"，因此波浪边必须位于侧面而不是顶部。过去，康沃尔郡当地锡矿工的妻子会在馅饼上标注丈夫姓名的首字母，防止矿工之间相互拿错。她们把馅饼做得十分结实，即使落入矿井也可以完好无损。

19 世纪，康沃尔郡当地的矿工大批地移居美国，同时为美国带去了他们的馅饼传统。在密歇根州上半岛地区和宾夕法尼亚州东部地区，这种传统至今仍然很流行，这也是美国少有的咸味馅饼传统之一。

[第七章]

虚构的派

然而，每晚宽衣解带，

虔诚地跪下膝来，

我都会在祈求上帝保佑的同时，

请求赐予我奶酪和苹果派。

——尤金·菲尔德

如果认为艺术和文学中展现给我们的派的盛宴几乎和从现实厨房中端出来的一样丰盛，那未免有些愚蠢。不过在那些构成我们文化遗产的故事和图片里，派的身影确实经常出现。我们在很小的时候就已经通过学唱儿歌接触到了派的意象，而那时我们甚至都还没有真正吃过派。要从这些明显很荒谬的儿歌里获得乐趣，并不一定需要理解它们。不过有些婴儿长大成为民俗学者或者都市传说学者之后，便确信自己小时候接触到的儿歌肯定具有特定含义，还坚定地认为这些含义会被发掘、揭示出来。有个事实阻碍了他们的探索，那就是歌谣的口述传唱早于书面记载，在某些情况下可能会出现长达几个世纪的时间差。因此，很多解释都难免包含大量的猜测。

有一些解释看起来想象力过于丰富，但它们通常可以分为两类：一类是基于历史事件和历史人物进行解读，也就是认为它们代表着大众媒体出现之前的流言或宣传；还有一类是把歌谣解释成教导儿童的警示性故事。《三只小猫》和《简单的西蒙》或许就属于后者。在《三只小猫》的故事里，小猫必须找到丢失的手套才能得到派；在《简单的西蒙》里，

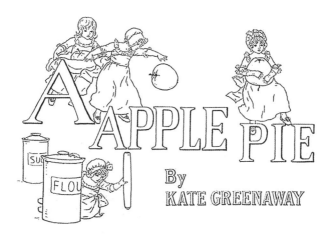

大多数儿童读物通常用字母 A 代表苹果。儿童读物插画师凯特·格里纳韦（1846—1901）在这里用字母 A 代表了苹果派。

西蒙虽然头脑简单，却在遇见卖派人之后懂得了买派不能赊账。

　　在《乔治·波吉、布丁和派》的故事中，乔治亲吻了女孩，使她们哭了起来。这里的乔治有两个可能的真实人物原型：第二任白金汉公爵乔治·维利尔斯（1628—1687）和曾担任摄政王的乔治四世（1762—1830）。这两个人都是出了名的不道德并且贪吃。据说《小杰克·霍纳》也是基于真实人物创作的，主角的真实姓名可能是托马斯。亨利八世解散修道院期间，托马斯曾经是格拉斯顿伯里教区修道院院长的管家。他从圣诞派里取出的"李子"[1]其实是价值不菲的萨默塞特郡梅尔斯庄园土地的地契。根据不同的故

────────────

① 童谣《小杰克·霍纳》："小杰克·霍纳坐在角落，吃着他的圣诞派。他把拇指伸进派，掏出一个李子便说道：'我真是个好孩子！'"——编注

事版本，可能是托马斯偷走了自己负责保管的地契，也可能这些地契是作为对其服务的酬报而赠予他的。至于故事提到的派是一个真实的派还是一种隐喻，读者完全可以进行自由猜测。毕竟我们已在前文看到，过去人们确实会在派里藏一些不可食用的东西。

审判红桃杰克。在路易斯卡罗尔的《爱丽丝梦游仙境》（1865年）中，红桃杰克因偷走红桃皇后的果酱挞而遭到指控。

多萝西·惠勒为童谣《小杰克·霍纳》创作的插画（1916 年）。

那么，《六便士之歌》里被烤成派的黑画眉又是怎么回事呢？对这首儿歌的解读至少有六个版本，其中一些听起来可能性非常小，还有一些听起来更是荒谬。不过，这个故事的起源可能要追溯到1454年。那一年，当时的勃艮第公爵，也就是"好人菲利普"在法国里尔市举办了一场宴会，借此博得人们对发动另一场战争的支持。宴会其中一个节目是在一个巨大的派里藏了一群音乐演奏家，派一打开就开始奏乐。据说当时其实有28名演奏家，不过数量上的微小差异并不会完全否定这个解释。还有一种观点认为，

从迪士尼动画片《鹅妈妈去好莱坞》（1938年）中截取的图像。这部作品用当时一些著名明星的漫画形象取代了歌谣里的人物。在这张图的场景中，奥利弗·哈迪成了童谣《头脑简单的西蒙遇到一个卖派人》里那个卖派的商人。

这首童谣可能与一天的时间有关，24只黑画眉代表了24个小时，而打开派和鸟儿唱歌代表破晓。某些都市传说坚称这首童谣是用来招募海盗的暗号，这种观点纯属无稽之谈。

当我们告别童年之后，自然会从童谣转向书籍和电影。这个时候，派仍然会化身为特定角色、情节推动因素、道具或者隐喻，出现在成人文学作品之中。在查尔斯·狄更斯的作品中，对派的运用和为派赋予的象征性意义简直可以写成一篇博士论文。狄更斯的派有时是一种热情好客的象征（比如《尼古拉斯·尼克尔贝》中当作早餐的巨大的约克郡派），却又常常带有迷惑性：《大卫·科波菲尔》中的鸽子派"像是一颗令人失望的脑袋，从颅相学的角度看，满是鼓包和肿块，底下什么特别的东西也没有"。有时候他的派又隐约流露出恶意，比如我们在第四章看到的《我们共同的朋友》里装满残羹剩饭的派皮。《匹克威克外传》中出现了一场狄更斯作品中令人最为舒心的有关派的讨论：

> "小牛肉派，"韦勒先生一边把食物摆在草地上，一边喃喃自语道，"小牛肉派是个好东西。只要你认识做派的女士，确定这派不是用小猫的肉做成的就好。不过话说回来，这种可能性又有多大呢？猫肉和小牛肉特别像，连那些卖派人自己都分不清了。"

接着，韦勒先生就开始复述他那卖派的朋友给出的建议：

"'关键就在于调味。派都是用那些高贵的动物做成的,'他说着,指了指一只非常可爱的小花猫,'我可以根据需要给它们调味,做成牛排、小牛肉或腰子。不仅如此,'他说,'我一旦得知市场发生变动或者人们口味有改变,就能把小牛肉做成牛排的味道,把牛排做成腰子的味道,或者把随便哪种肉做成羊肉的味道!'"

　　在弗朗西斯·霍奇森·伯内特的作品《小公主》中,一个小小的肉派所提供的并不仅仅是基本的营养。在故事中,莎拉小姐和营养不良且劳累过度的小女仆贝基之间产生了一种不合身份的友谊。善良的莎拉小姐经常会找来一些小点心,偷偷见面的时候塞给贝基。有一次,莎拉小姐带来了一些小肉派,这对于小女仆来说是一份让人惊喜的佳肴。贝基非常兴奋:"这真是填饱肚子的好东西。填饱肚子最重要。海绵蛋糕也能填饱肚子,但不知小姐是否知道,它们在肚子里化掉以后会堵在胃里。"吃过莎拉小姐带来的食物之后,贝基的健康状况明显出现了好转,但是支撑贝基的东西不仅仅只是食物。她知道,"就算没有肉派,单单能够见到莎拉小姐就已经足够了"。

　　诗人路易斯·昂特迈耶曾经产生过一种疑问:"为什么我们的诗歌总是回避对食物的着迷和回应?"诗歌并没有完全回避这个主题,但似乎确实把真正的抒情诗都留给了那些甜美的水果;或者偶尔出现像巴勃罗·聂鲁达那样的诗人，把辞藻和情感献给了番茄、洋葱等蔬菜。有一些诗歌涉及了派，但是倘若它们不是滑稽类作品，那么真正描绘的通常是更大的主题。罗伯特·骚塞的《醋栗派》总的来讲是一首对造物的赞歌，也是一首特别献给"简"的颂诗；在约翰·格林利夫·惠蒂尔的诗歌作品《南瓜派》中，南瓜派是感恩节的象征性符号；本章节开头的那段知名引语其实选自"儿童诗人"尤金·菲尔德为激发爱国热情所创作的一首诗歌。

可视媒体出现之后，派又有了其他大放异彩的机会。虽然派可能没有频繁作为主要角色出现在电影里，然而一旦出现，就会十分抢眼。谁能忘记电影《美国派》中那热乎乎的自制苹果派呢？这大概是有史以来唯一一部把派当作性对象的作品。

有一部1901年的早期默片名叫《热羊肉派》，这部电影有意利用了民族焦虑和复仇心态：两个男人从一个中国小贩手中买了一个派并且吃了下去，而那个小贩却令人毛骨悚然地笑了起来，他把牌子转过来，上面写着"都是猫肉派"几个字。食客对于派馅料的忧虑是文艺作品反复使用的主题。由文森特·普赖斯主演的恐怖电影《血染莎剧场》（1973年）对这一主题进行了更加生动的刻画。在影片中，普赖斯扮演了一名遭遇评论家批评的莎士比亚戏剧演员，为了报复，该角色根据每个评论家对他的评论展开了针对性的谋杀。在其中的一幕场景中，评论家梅雷迪思·梅里杜（罗伯特·莫利饰演）因被迫吃下用自己"孩

《美国派》（1999年）这部
电影首次使苹果派成了性的
象征。

《李阿姨的肉派》（1992年）
同样讲述了把人肉做成派、卖
给不知情食客的故事。

子们"（他心爱的贵宾犬）的肉做成的派而噎死；这一幕
显然受到了莎士比亚作品《圣诗复仇》的启发。人类对
于同类相食的现象有着更深的（几近原始的）恐惧和迷
恋，而电影《李阿姨的肉派》（1992年）对这种心理进行了
进一步描绘。这部电影的内容并不新颖，直接取材于18世
纪的"理发师陶德"事件。在电影中，李阿姨的派十分有
名，都是在她四个漂亮侄女的协助下制成的。侄女们还会
帮她找寻馅料的制作原料——英俊的年轻男子。

　　在讨论电影中的派时，不得不提一种经典的喜剧手
法——扔蛋奶派。"拍派"现今已被合法化，成为一种半严
肃的颠覆性政治行为。为了表达自己的观点，人们把奶油

电影《世纪之战》（1927年）中出现了电影史上规模最大的派大战。

"拼命阿丹"狼吞虎咽地吃牛派。

在电影《女招待》（2007年）中，一个具有烘焙天赋但生活不幸福的女服务员通过开发新型的派来发泄自己的情绪，最终借此获得了成功与幸福。

网络动画角色韦伯和鲍勃，以及它们喜爱的派。

派送（"满含情意地推动"，而非投掷）到相应的名人的脸上，而且最好是在全球媒体的眼皮子底下——这种行为就是"拍派"。1984年，著名的比利时无政府主义者诺埃尔·戈丁亲手把派拍到了比尔·盖茨的脸上。戈丁认为，这种行为是一种交流方式——"一种视觉上的世界语"。对于目标人物和旁观者来说，有时"拍派"实际传达的信息并不清晰；不过那些声称自己受到喜剧电影扔派场景启发的肇事者大概并不这么认为。喜剧组合"劳雷尔与哈迪"的电影《世纪之战》中有一组长达四分钟的镜头，堪称电影史上最壮观的派之战。对于那些准备成为"专业拍派者"的人来说，这段画面或许是一堂必修课。在《世纪之战》的拍摄过程中，剧组订购了3000多个奶油派（相当于洛杉矶派公司一整天的产量），并全部用完。在场所有人员都在向自己的目标扔派，每个人都玩得很开心。

最后，让我们把目光转向卡通动画。很多卡通角色都有自己最喜欢的食物，比如加菲猫喜欢意大利千层面，达格伍德喜欢大三明治，吉格斯喜欢咸牛肉和卷心菜。在有些动画中，角色最爱的食物具有推动故事情节的重要作用，比如大力水手的菠菜。"拼命阿丹"是英国漫画《丹迪》中的一名超级牛仔英雄，自1937年起就被称为"世界上最强壮的男人"。拼命阿丹力气非常大，而他的力量源泉就是"牛派"——你没听错，这种派里塞了一整头牛，包括牛角、牛尾等所有部位。遗憾的是，70年之后，为了防止人们误认为这个人物在宣扬疯牛病，他的力量之源便不再出现了。

在动画领域，最突出的派爱好者要数 Flash 动画剧集《韦伯的日常》中的主要角色。这部有着一批狂热粉丝的动画在各方面都践行了极简主义：韦伯和他的朋友鲍勃就是两个鸡蛋形状的动画人物，总是在洋红色的背景上滚来滚去，伴着音乐说着简短的对白。它们的极简日常生活常常围绕着寻找派展开。

派的未来

> 有了那些预拌食品，你可以认为下一代人做派的时候再也不用像妈妈们过去那样处理一些麻烦事。
>
> ——厄尔·威尔逊

我在本书开头提出过一个问题："派会消失吗？"今天的我们已经拥有了烤盘、午餐盒和冰箱等工具，可以用更好的方式实现派最初的那些功能。我们不再"需要"派了。那么我们是否还会"想要"派呢？

商业化的派似乎勉强存活了下来，仍然在以老牌街头小吃的身份与热狗、土耳其烤肉等同类型食品进行竞争；不过今天的人们仍然会像乔叟那个时代的人们那样，担心派的馅料问题。有些派（比如圣诞节百果派和感恩节南瓜派）之所以很好地保留了下来，是因为这些派已成为一种强有力的象征性符号，不再只是美味的卡路里。至于那些普通的家庭自制派，它们的命运则凄惨很多。在我们目前这个快节奏且忌讳油酥面皮的世界里，派还能走多远呢？做派是一件苦差事，或者可以说是一件充满爱意的苦差事，而今天的我们并没有什么时间下厨。另一方面，现代生活十分便利，我们完全可以去买冷冻油酥面皮，甚至可以在台式烤派机[1]里做派，连烤箱也用不到。我们会不会懒到连烤派机也用不上呢？

---

① 类似电饼铛的厨房家用电器，上下盘有若干圆形凹槽以贴合派的形状。——编注

现代做法的圣诞节百果派只用到了水果，不像中世纪时期，人们把肉和水果混合起来作为这种派的馅料。

批量生产的商业化的派

感恩节派

如果派的世界真的像我现在担忧的那样正在分崩离析，那会不会产生任何影响呢？会有人在意吗？我们是应该单纯地带着怀旧之情回顾过往，记录派的消逝，还是应该尝试挽救派呢？通常来讲，像这样的问题还会引发更多其他问题。拯救海豹的同时无法做到拯救海豹鳍派，但如果制作海豹鳍派不必残忍杀害海豹幼崽，那么海豹鳍派是否值得我们去保护呢？尝试拯救光芒四射的约克郡圣诞节派的同时，能否忽略血派和马肉派呢？就我个人而言，我肯定不想去拯救午餐肉派，但其他人可能持有不同观点。

优秀的派不仅仅是烹饪艺术的杰出典范，同时也是整个人类烹饪历史的"一道"缩影。我们应当救下它们，不是吗？

食

谱

油酥面皮制作是烹饪的诸多领域之一，并且要求测量精确。早在 1615 年，杰维斯·马克海姆①就已经指出，厨师本身就应该知道"哪种肉适合使用哪种油酥面皮"。因此并非所有种类的派都有油酥面皮制作方面的食谱。

到了现代，厨师只要掌握了一份基础的简易油酥面皮的食谱就能制作大多数的派。一小部分热衷于派的爱好者可能会亲手制作千层酥皮，不过商店里出售的优质冷冻油酥面皮也完全适用。出于一些显而易见的原因，几乎没人会去亲手制作费罗酥皮。不过，所有人都应该至少尝试制作一次烫面酥皮——烫面酥皮的制作与其他油酥面皮相比让人省心得多。

基础的简易油酥面皮

基本配方为面粉加上重量为面粉一半的动物油脂，加足够的水把面粉和油脂混合在一起。过程中注意让所有食材保持低温，尽量不要用手接触。以下用量足够做出 8 英寸（约 20 厘米）的双层皮派。

原料：

8 盎司或 220 克未加入发酵粉的普通面粉，过筛；

少许盐；

4 盎司或 110 克动物油脂（黄油或者黄油和猪油的

---

① 杰维斯·马克海姆（Gervase Markham，约 1568—1637），英国诗人、作家，著有烹饪食谱书《英国主妇》。——编注

混合油脂）；

2~4 汤匙冷水。

面粉加盐，把油脂揉进面粉里（也可以使用搅拌机简单地搅拌一下），呈现类似面包屑的状态即可。加水，然后用刀片快速地轻轻搅拌。当面团不再粘连碗壁的时候，用手指把面团轻轻拍实，再用保鲜膜把面团包起来，放进冰箱冷藏至少半小时。条件允许的话，可以在烘焙之前把擀完的面团再冷藏半小时。

烫面酥皮

适用于比较结实的肉派或馅饼。以下用量可以做出 4 个大号馅饼。

原料：

10 盎司或 400 克普通面粉加一大撮盐，过筛；

4½ 盎司或 125 克猪油或烤肉流出的油滴（可将一

半替换成黄油以调味）；

5 盎司或 150 毫升牛奶或加了水的牛奶。

把牛奶和油脂放入锅中煮沸，加入面粉搅拌至表面光滑平整。把面团切成 4 份，趁热擀成圆形面皮（厚度约 ¼ 英寸或 ½ 厘米）。根据需求填充馅料。

康沃尔馅饼

原料：

1 磅或 450 克面粉制作而成的简易油酥面皮或其他

类型油酥面皮面团；

¾ 磅或 340 克牛肉，切成小块（不要剁碎）；

1 个洋葱，切碎；

1 片生洋蔓菁或生萝卜，切成小块；

2~3 个生土豆，切成小块；

1 个打散的鸡蛋，上光用（可选）。

把面团擀成厚度约 ¼ 英寸的面皮，切成若干个圆形，尺寸自定。把其他原料轻轻混合。想要沿面皮边缘捏合的话，就把混合的原料堆在半边面皮上；想要在派的顶部捏合的话，就把原料放在面皮的中央位置。面皮边缘蘸水，捏合成波浪纹。想要派的表面有光泽，可以刷上打散的鸡蛋液。送入烤箱，以中火烤 35~45 分钟。

仰望星空派

原料：

6~8 条沙丁鱼；

1 个中等大小的洋葱；

3 片培根薄片；

1个柠檬；

2个散养鸡下的蛋；

18盎司或500克简易油酥面皮或香脆酥皮[①]面团；

盐和胡椒，调味用（使用海盐风味更正宗）；

欧芹和龙蒿，调味用。

康沃尔式做法

清除沙丁鱼的内脏和鱼骨，保留鱼头和鱼尾（新鲜的沙丁鱼可以在不用刀的情况下拔除脊骨）。洋葱切碎，培根切成小方块。把柠檬对半切开，切两片放在派的一侧进行装饰，剩余柠檬挤出柠檬汁暂时存放，磨碎柠檬皮。把鸡蛋煮熟，切成小块。

把油酥面皮面团切成两半，擀成面皮，其中一张面皮放入8英寸的浅底派盘里。切掉露出盘子边缘的面皮，面皮边缘涂抹牛奶或水，以确保油酥面皮顶面可以黏合。

做法一：把沙丁鱼摆放到盘子底部，沿盘子边缘呈车轮辐条状排列。把事先切碎的洋葱、鸡蛋以及培根摆放到沙丁鱼之间的空隙中。有一些食谱会建议把一半切碎的配料填充到鱼肚子里，但考虑到沙丁鱼腹部空间较小，可以不做填充。倒入柠檬汁，盖上油酥面皮顶，沿沙丁鱼周围压实油酥面皮封起来。切掉两面油酥面皮重叠后边缘多出来的部分，捏出正宗康沃尔风格的波浪状边缘。

---

① 香脆酥皮酥脆如千层酥皮，但没有明显分层，是千层酥皮的简易版本。——编注

做法二（这种做法更正宗）：把所有切碎的配料（包括调味料）放进盘子里。同上，盖上油酥面片顶，把边缘修整齐，捏成波浪状。用刀小心地在油酥面片顶表面切出口子，用刀刃撑开切口，轻轻塞入整条的沙丁鱼，露出鱼头和鱼尾。加入柠檬汁后封住切口，最后把打散的鸡蛋液涂抹在派的表面。

烤制

把烤箱预热到200℃，把派放入中央，烘烤大约30分钟，烤至表面呈金褐色（尺寸更大的派需要烤更长时间）。摆上一小枝欧芹进行装饰，再配上康沃尔当地的新鲜土豆，趁热端上桌。

历史食谱

如果一个现代厨师尝试去重现历史食谱中的菜肴，那他可能会深感挫败。那些古老的烹饪书很难作为现代的操作指南，因为这些书里面很少写明具体的数量、温度等信息，内容非常含糊。烹饪书的年代越久，这种问题也就越严重，而且古时候的语言常常令人费解。中世纪时期的烹饪书并非是按照操作指南来写的——当时的厨师都是在工作的过程中学习，不需要书籍，而且不一定具备读写能力。这些古老的烹饪书其实是作为备忘录供那些负责为家里采购的人使用，或者是放在主人的书房里用来彰显主人的财富。直到

19 世纪中期，烹饪书里的操作指南才开始变得更详细，更精确。对于大多数历史食谱，我们都需要有根据地去推测，并运用自己的判断。现在请感受一下下列历史食谱：

布莱挞
——选自《烹饪之法》（约 1390 年）
作者：英王理查二世的主厨团队

在几英寸深的容器里放上油酥面皮。准备艾伦地区产的生鸡蛋黄和卢恩奶酪，把奶酪融化与蛋黄混合在一起，然后添加姜粉、糖、藏红花和盐。把容器放进烤炉进行烘烤，烤完取出。

注："布莱"可能指的是这种奶酪挞的原产地。特别提到的"卢恩奶酪"是一种成熟过程中经多次水洗、表面光滑的软奶酪。这份食谱里没有提到糖的具体用量，因此我们无法得知这个挞的甜度。不过糖在当时非常昂贵，由此可以推测这道菜品大概更接近于今天的法式咸挞，而不是芝士蛋糕。

苹果橙皮挞
——选自《好厨娘手册》（1597 年）
作者不详

把橙子放进水里浸泡一天一夜后用蜂蜜水煮，煮软之后再把橙子放到糖浆里浸泡一天一夜，捞出橙子切成小块备用。准备好挞皮，使用糖、肉桂、姜给苹果调味。在挞里放一片黄油，铺上一层苹果，再在苹果的间隔处铺上一层橙子，如此反复。怎么给苹果调味，就怎么给橙子调味，不过要放更多的糖。盖上油酥面皮放入烤炉，快烤好的时候把玫瑰水加糖煮至浓稠。把挞从烤炉里取出来，插上一根羽毛，把煮过的玫瑰水涂到油酥面皮顶面。注意不要烤焦。

注：这是一道非常奢侈的菜品。第一步提到的"糖浆腌橙子"（"橙皮蜜饯"）在当时是一种非常昂贵的美食。1600年，伊丽莎白一世的两名糕点主厨各自做了一个橙皮蜜饯派献给女王作为新年礼物，由此可见像橙皮蜜饯派这类派在当时是多么高贵。

大斋节香草派

——选自《资深英国管家》（1769 年）

作者：伊丽莎白·拉法尔德

取生菜、韭葱、菠菜、甜菜、欧芹各一把，煮过之后切成小片。用一块布把 1 夸脱[1]去壳谷粒包起来，加两三个洋葱一起煮熟。把所有材料放进煎锅里，加入香草、大量的盐、1 磅黄油和少量切成薄片的苹果炖几分钟，随后就可

---

[1] 1 夸脱约等于 1.14 升。——译注

以装入盘子或者油酥面皮进行烘烤。烤 1 小时即可上桌。

南瓜派

—— 选自《美国烹饪》（1796 年）

作者：阿梅莉亚·西蒙斯

做法一：准备 1 夸脱炖熟且滤过水的南瓜、3 品脱牛奶、6 个打散的鸡蛋、糖、肉豆蔻衣、肉豆蔻、姜，放进 7 号或 3 号油酥面皮中，摆成十字方格状，然后放入盘子烘烤 45 分钟。

做法二：准备 1 夸脱牛奶、1 品脱南瓜、4 个鸡蛋、甜胡椒、姜，放入油酥面皮烘烤 1 小时。

7 号油酥面皮（"适用于甜肉的油酥面皮"）

准备 2 磅面粉，加入 ⅓ 磅黄油和 1 磅猪油，用 4 个打散的蛋清拌湿，加入足量的水做成面团。将残留的起酥油分 10~12 次揉进去之后快速烘烤。

注：这是美国印刷的第一本烹饪书，上述食谱出现在"布丁"一章。书中还有一份"马铃薯派"食谱，实际上用的是甘薯。

咖喱派（鱼馅或者肉馅）

——选自《家政与烹饪，穷富皆适用》(1827 年)

作者：一位女士

使用现成的已经冷却了的咖喱，最好使用过夜咖喱。不管什么时候剩下的咖喱都要比新做的咖喱效果好。把咖喱放进一块优质千层酥皮里，顶面盖一层薄酥皮。用一些长叶片沿着顶面边缘紧紧围一个圆，插的时候叶尖朝上，让叶片从高高的顶上一直垂到底部，这样的造型会十分美观。

咖喱派须搭配咖喱饭，或者配了咖喱酱的普通米饭。

美味的苹果挞

——选自《现代烹饪大全》(1845 年)

作者：伊莉莎·阿克顿

1¼ 磅去皮去核的苹果足以做成一个小型的苹果挞；再添加 4 盎司苹果就可以制成中等尺寸的苹果挞。盘子里放一张英式千层酥皮或者奶油酥皮，把苹果蘸水之后紧密排列在酥皮里，中央部位的苹果需要比四周高一些。在苹果中撒上 3~4 盎司捣碎的糖，如果苹果非常酸，则增加糖的分量。取半个柠檬，把磨碎的柠檬皮和挤出的柠檬汁添加到苹果里，可改善口味。盖上擀薄了的酥皮顶，加不加糖霜都可以。把苹果挞放进中等火力的烤炉烘烤大约半个小时。我们也可以把这款苹果挞做成老式的奶油苹果挞：趁热将酥皮顶切掉，只在边缘留下 1 英寸宽的酥皮边，苹果变冷之后倒上 ½～¾ 品脱煮开的浓郁蛋奶酱。过去人们会把酥皮顶切分成三角形的小片插到苹果挞里面，不过用泛白的千层酥皮薄片作为点缀效果会更好。可以使用无多余水分的掼奶油代替蛋奶酱，奶油应该堆得高一些，轻轻盖在苹果上。

苹果卷饼

——选自《国际犹太烹饪书》（1919 年）

作者：弗洛伦斯·克赖斯勒·格林鲍姆

准备一个大碗，放入半杯面粉和 ¼ 茶匙盐。轻轻打散一个鸡蛋，加入 ⅓ 杯温水，和面粉与盐混合。用刀快速搅拌面粉，形成面团，然后放到砧板上上下拉伸增强弹性，直到面团不粘连砧板为止。把揉好的面团放到一块撒了面粉的砧板上，盖上一个热碗，在温暖的地方放置一段时间。准备馅料的同时把面团放在撒了面粉的桌布中央，把面团略微擀开，刷上融化的黄油，然后把双手放在擀开的面团下面，手掌向下，轻轻拉伸，直到面团像桌子那么大，像纸一样薄。这里注意不要撕坏面团。准备 1 夸脱去皮切碎的酸苹果，¼ 磅热水焯过后切碎的杏仁，还有半杯麝香葡萄干和黑科林斯葡萄干、1 杯糖、1 茶匙肉桂，均匀地铺在 ¾ 的面皮上，洒上几汤匙融化的黄油。修整面皮边缘，然后把面皮朝着苹果馅料的方向卷过去，双手把桌布拉高，这个卷饼就会自己不停地滚动，卷成一个大卷。再次修整边缘。扭一扭这个卷饼，根据涂了油的平底锅尺寸调整它的尺寸，放入烤炉烤至褐色且酥脆之后，在表面刷上融化的黄油。如果使用多汁的小型水果或者浆果作为馅料，就要在拉伸之后的面团上撒面包屑吸收汁液。冷却至温热程度即可端上桌。

注：苹果卷饼能不能算作派呢？当然可以。它虽然没有放在特定形状的容器里，但同样有一层油酥面皮包裹着，完全符合派的特征，不是吗？

牛排腰子派
—— 选自《节俭的厨师》（1851 年）

作者：E. 卡特

充分敲打牛排，让牛肉的口感更嫩。添加重量为牛肉三分之一的腰子。腰子要切成小块，这样有利于烤出全部的肉汁；使用胡椒和盐进行调味。沿着盘子的侧面和边缘铺上面皮，再用油酥面皮把整个派盖起来。根据肉派的做法对牛排腰子派进行装饰。

注：几个世纪以前人们就开始使用肉和腰子制作派，不过直到 19 世纪晚期，"牛排腰子派"这种说法才变得常见。比

顿夫人于 1861 年给出了一份牛排腰子布丁的食谱，不过她的基本款牛排派只建议添加牡蛎、蘑菇和洋葱末作为配料。几个世纪以来，牡蛎一直都是牛肉派的传统配料，并且在比顿夫人的时代，牡蛎是穷人也消费得起的廉价食物。随着维多利亚时代的终结，牡蛎变得昂贵起来，腰子有可能就是在这个时候开始成为牡蛎的常见替代品。

葡萄干小羊肉派
—— 选自《卡斯尔烹饪辞典》（约 1875 年）

取 2 磅小羊胸肉，切成整齐的小块放进派烤盘里。在羊肉表面撒上 1 餐匙盐、1 茶匙胡椒、1 汤匙欧芹末、¼ 个磨碎的肉豆蔻、3 汤匙挑选过的黑科林斯葡萄干。把两个鸡蛋完全打散，添加一葡萄酒杯的雪莉酒，然后一起倒在羊肉上。沿着盘子铺一层质量上好的油酥面皮，顶上再盖一层，放入火力中等的烤炉烘烤一个半小时，端上桌的时候要搭配一点白葡萄酒和糖。除去葡萄酒之外，成本大约为 2 先令 8 便士。这个派足够四五个人食用。

注：混合果干和小羊肉，食用的时候搭配糖和葡萄酒，这种烹饪和食用的方法可以追溯到中世纪时期。

杏仁小挞

—— 选自《实用烹饪百科全书》（约 1891 年）

作者：西奥多·弗朗西斯·加勒特

准备一打挞模具，沿着模具边缘铺上面皮，然后在底部涂上薄薄一层柑橘果酱。取 6 盎司热水焯过的杏仁，放入烤炉烘干，然后添加等量的细砂糖，加上一点橙子皮或者柠檬皮，和 6 个蛋黄搅拌在一起。把混合后的馅料从研钵倒进盆里，加上 8 个打发的蛋清搅拌均匀，然后填充到小挞里。填满之后表面撒上一些细砂糖，送入没有预热的烤炉烘烤 25~30 分钟。

乳鸽锅派（英国风格）

—— 选自《圣弗朗西斯酒店烹饪书》（约 1919 年）

作者：维克托·希尔兹勒

取若干只乳鸽烤熟，切成两半。使用融化的黄油把一片牛肉的两面迅速煎一下。把红葱头切碎，煎牛肉片的时候一起翻炒，然后把乳鸽、牛肉片、红葱头一起放进派烤盘里，再添加 6 个罐头蘑菇或者新鲜蘑菇、半个大火煮熟的鸡蛋、少量切碎的欧芹和加了面粉的烤乳鸽肉汁，然后加入少量伍斯特郡酱充分调味。在盘子上覆盖一张油酥面皮，烘烤 20 分钟。这是一人份的派，如果需要制作更大尺寸的派，可以根据比例增加原料。

西梅干葡萄干派

—— 选自《国际犹太烹饪书》（1919 年）

作者：弗洛伦斯·克赖斯勒·格林鲍姆

取半磅西梅干，煮软至肉核分离，捞出后用叉子捣碎，再加入煮时的汤水。取半杯麝香葡萄干，用少量的水煮几分钟，煮软后连同半杯糖一起添加到西梅汤中。少量丁香粉或者柠檬汁可以提升口味。烘烤的时候要用两张派皮。

柠檬奶油派

—— 选自《优雅的烹饪艺术》（1925 年）

作者：C.F. 勒韦尔夫人，奥尔加·哈特利女士

取 1 杯糖、1 杯水、1 个磨碎的生土豆，再取 1 个柠檬挤出柠檬汁、磨碎柠檬皮。把所有原料混合起来，上下两面分别包上油酥面皮后进行烘烤。

鸡蛋培根派

—— 选自《厨房子爵》（1933 年）

作者：莫杜伊子爵

这是一道非常棒的约克郡美食。

在派烤盘里铺上香脆酥皮面皮。底部铺上两层瘦培根，培根上轻轻打三个鸡蛋，加入盐和胡椒，把无覆盖的烤盘

送入烤炉烘烤。鸡蛋凝固之后在鸡蛋上方再铺两层培根，培根上再打三个鸡蛋，添加胡椒。用面皮把派盖起来，表面涂上鸡蛋液后送入高温烤炉，酥皮烤熟后取出。

蔬菜派

——选自《366 份菜单和 1200 份食谱》（1868 年于法国首次出版，此处选用的是 1896 年的英语版本）

作者：布里斯男爵

取适量青豆、嫩蚕豆、小胡萝卜、嫩四季豆，分别用奶油酱煮熟。在烘烤过的派皮里用油酥面皮薄片分隔出若干个区域，把上述蔬菜分别放进去就可以端上桌了。冬天可以使用腌制的蔬菜来制作。感谢声名卓著的格里莫·德·拉雷尼耶发明了这道美食。这道蔬菜派不仅味道好，而且赏心悦目，也为斋戒日的餐桌增添了一份美味。

致

谢

本书的创作过程充满乐趣。很高兴能得到这次机会来针对本书主题进行深入的研究与写作，该主题对于我个人的心灵与传承来说非常珍贵。同时很荣幸能够与诸多优秀作家一起参与"食物小传"系列图书的工作。

首先感谢烹饪历史学家兼作家安德鲁·F. 史密斯推荐我参与该系列图书项目，感谢瑞科图书出版社的迈克尔·利曼接受安德鲁的推荐，还要感谢玛莎·杰伊在后期创作阶段对我的耐心帮助。

特别感谢我的女婿帕特里克·布赖登，他给予了我无价的帮助，本书中的许多图片都是他搜集准备的；感谢我的丈夫布赖恩和我们特别的朋友特雷弗·纽曼，他们努力忍住了食欲，给梅尔顿·莫布雷派和佩兹纳斯派拍了照片。同样感谢我的博客"老美食家"的各位忠实读者，他们帮助我坚定了这一观点：除历史学家以外，"普通读者"同样会对食物的历史感兴趣。

最后，无比感谢我的家人和朋友一直以来的支持与不懈的热情，特别是我的丈夫布赖恩，我的孩子马修、莎拉以及他们的伴侣维基、帕特里克，还有我的妹妹瓦珥。衷心地感谢各位。

Pie

A Global History

Janet Clarkson

Contents

Prologue: Preliminary Observations on Pie

I may not be able to define a pie, but I know one when I see it.

Raymond Sokolov

It behoves, I am told, every author to consider the scope of her work before launching into it. No problem, I thought. It is simple. It is about Pie. It is also, I am told, essential for the author to describe the scope of her work at its beginning, in order to forewarn her readers. Simple, I thought. I will start by defining Pie. In the end, I failed utterly. I would have agonized less over my failure had I remembered the above quotation rather earlier in the piece than I did. If the illustrious and erudite Raymond Sokolov cannot define pie, who am I even to attempt it? Nevertheless, I don't regret the attempt, as it was fascinating and enlightening in ways that I would not have suspected.

I began with a memory I had of a venison pie I once ordered at an upmarket restaurant. After exactly the right anticipatory interval my dinner arrived. It was very elegant. On a base of fluffy garlic mashed potatoes was a cleverly layered construction of julienned vegetables topped with slivers of tender, rare venison, the whole surmounted with a precariously balanced disc of puff pastry. It was delicious, but it was not pie. I was quite sure that it was not pie. Clearly, the chef and I had vastly differing opinions on the issue, which brought me to an early epiphany. When—and if— I came up with my concise, accurate definition of pie, not everyone else in the world would agree with it.

But surely there would be some areas of consensus as a starting point? Pies are not pies simply because they are called pies. The American treat called Eskimo pie, for example, is unequivocally ice cream, and moon pie is a chocolate biscuit with pretensions. Boston cream pie is certainly a cake, but apparently called pie because it is baked in a pie tin, by which logic I could cook porridge in a pie pan and call it 'porridge pie'. Other 'pies' are more problematic. Is a sausage roll a small pie? Should I include cobbler and pandowdy, which seem like failed pies with broken lids? Traditional Scottish black bun and English simnel cake are fruit cakes with pastry shells. Do they count? I pondered cottage pie and shepherd's pie. These have a layered pie-like structure certainly, but were (to me) instantly definable as not 'real' pies. Why not? Another epiphany dawned. Because they have no pastry. It seemed that my attempt at definition was devolving into a set of minimal criteria, at least for the purposes of this book. This second epiphany led to the First Law of Pies: 'No Pastry, No Pie'.

For reasons which will become apparent in chapter One, I immediately came up with the Second Law of Pies:

they must be baked, not fried (or boiled, or steamed). One more law as to the number and location of crusts required (single top, single bottom, or double), and my criteria would be set—or so I thought. Formulating the Third Law of Pies proved to be an extraordinarily difficult part of the challenge, and I am still far from sure that I have resolved it in a satisfactory manner. I began my crust count with the *Oxford English Dictionary*, which asserts that a pie is:

> A baked dish of fruit, meat, fish or vegetables, covered with pastry (or a similar substance) and freq. also having a base and sides of pastry. Also (chiefly *N. Amer.*): a baked open pastry case filled with fruit; a tart or flan.

It would seem then, that the editors of the *Oxford English Dictionary* are of the opinion that, to qualify as a pie, the top crust is essential and the bottom crust optional, except in America where it is the bottom crust that has primacy. They also appear to be suggesting that the bottom-crust American pie is, according to British-English usage, a tart or flan. The definition becomes less clear when they go on to describe a flan as 'an open tart', and a tart as 'the same, or nearly the same as a pie'.

I, who love and revere the *Oxford English Dictionary*, have to admit that it displays a lack of clarity in respect of pies and tarts, as indeed does most of the English-speaking world. Nowhere has this been more passionately demonstrated than in an extraordinary debate in the correspondence pages of *The Times* in 1927. In September of that year, a Mr R. A. Walker was moved to write a letter of 'vigorous protest' on 'the abominable soul-slaughtering and horrible trick of serving puff pastry and stewed fruit under the guise of apple tart', for which he blamed restaurateurs. Mr Walker's issue was to do with pie quality, but he was quickly taken to task by Colonel John C. Somerville for using 'apple pie' and 'apple tart' as alternatives. The Colonel could not 'for one moment allow' this, remarking that 'all properly brought up children' would know the difference. He went on to define 'pie' as 'whatever is cooked in a pie dish under a pastry roof', whereas 'when the fruit lies exposed in a flat substratum of pastry, then, and only then, can it be rightly called a tart'. A surprisingly passionate debate ensued over the next two weeks, with veiled insults and categorical statements being made and side issues being thrown in. Several expatriate American and French readers weighed in with their opinions and, as the discussion was becoming rather heated, 'an advocate of concise nomenclature' appeared in the form of Professor Henry E. Armstrong. The professor 'heartily supported' the discussion, and proceeded to give, as his best example, 'rice pie, always with an emphasis on pie'. He described 'rice, with a due amount of sugar and a plentitude of farmhouse milk' baked 'in a pie-dish; therefore a pie, the more because—in the jargon of the scientific—it is delicately auto-encrusted'. He clarified that this dish was not pudding, because 'a pudding must be boiled in a cloth and have a crust, but without sign of brown'. The professor's academic field was not indicated, but his slightly pompous opinion precipitated an indignant response from Mr E. E. Newton, who called his expertise into question, declaring that 'he cannot be a professor of cookery'. Mr Newton pointed out that the 'crust' of the professor's pie was simply the skin formed by the heat of the oven, whereas a pie has a 'crust formed of something different from the contents'. Another offended correspondent asked 'the stickler' how he would classify the famous Kentish 'pudding-pie' with its 'substratum and deep vertical wall of pastry . . . filled to the brim with custard and cooked in a container in an oven' and warned: 'If you want this dish in Kent, don't ask for custard tart.'

The Times staff ran a short editorial in the middle of all of this hearty support and indignant rebuttal, for the sake of 'all who hold that a pie is not a tart and a tart is not a pie'. The article acknowledged that the 'great Oxford Dictionary' and Mr Fowler's book on modern English usage 'throw up the sponge' in the matter of pie and tart, and it also gently chided Professor Armstrong for bringing puddings into the debate in an unhelpful manner. The discussion was to receive some 'authoritative support' in the form of an opinion from the manager of that bastion of English cuisine, Simpson's on the Strand. Mr F. W. Heck was unexpectedly humble. He did not want to add unduly to the correspondence, making 'a suggestion, or rather a plea, that English people should do their best to combat the tendency there seems to be these days to re-dress old English words with new meanings'. He invoked Charles Dickens, saying that that gentleman would have had no doubt what he meant by pie, and he would have meant, and got, 'that delicacy, made in a pie-dish, the interior having been cooked together with and under its golden pastry cover'.

Another authority in the unlikely form of Mr Gladstone was quoted a week later by Mrs Hugh Wyndham, who had heard him 'declaim on the subject with fervour and elegance'. He apparently contended that 'pie should be used only for fruit, that for meat, pasty was the right word'. It might have been expected that this would open up a whole new discussion on the pie/pasty issue but perhaps *The Times* readers were getting bored with the topic, for there were no responses. A final letter came from Mr Frank Birdwood, who stated with great confidence that the words 'tart' and 'torture' spring from the same Latin root and that 'it is certainly the case that the tart gets its name from the twisted (i.e. tortured) lengths of pastry with which the jam, or other principal ingredient, is still usually decorated, leaving it in all other respects "open".'

The 'gay battle of the tarts and pies' was formally ended on 12 October by an editorial piece announcing the commencement of the official 'pudding season'. The editors seemed relieved to note the copious consumption of puddings (sweet and savoury) by the English, this apparently indicating that: 'At the heart of the Empire, at any rate, our ancient stamina is not altogether lost.'

The professor who extolled the virtues of rice pie in *The Times'* debate could have found support for his argument in a decision made by the Supreme Court of America in 1910 that rice pudding was pastry. It all began when John Mylonopoulos rented the front part of a store to John Cobatos, one of the conditions being that the latter would not sell several specific items including 'pastry of any sort'. Some time later, Mr Cobatos commenced selling both prunes and rice pudding, managing close to 3,000 plates of the pudding in the next 47 days. Mr Mylonopoulos sued. The first judge crossed prunes off the claim immediately, dropping them 'into the respectable obscurity which is their humble but happy history'. The rice pudding issue was sent to a jury, which decided that it was pastry. Naturally there was an appeal. In spite of Mr Cobatos's counsel's 'fiery oration', quotations from many dictionaries and discussion of common understanding of the nature of rice pudding, the appeal was lost. Plaintiff Mylonopoulos was awarded his $101.05 damages and $21.50 costs.

If the *Oxford English Dictionary*, the erudite nineteenth-century readers of *The Times* and the Supreme Court of the United States fail to reach consensus on the question, 'What, exactly, is a pie?', then for the purposes of this book, I feel free to make up my own rules on the crust issue. Historically, one of the seminal features of a pie is its ability to be eaten out of the hand, so this would seem to exclude pot-pies. I leaned towards this restriction for a while, but common sense and common usage prevailed, and the proposed Third Law became redundant: any number of crusts would qualify, but these must (as stated in the First Law) be of pastry. Pastry, not a 'similar substance'—a decision based, I believe, on sound historic reasons which will be revealed in chapter One.

Words often give clues to the origin of things, and I hoped to learn much about the history of the pie from the word 'pie'. The *Oxford English Dictionary* gives its first known use as being in the expense accounts of the Bolton Priory in Yorkshire in 1303 (although the surname 'Pyman' is recorded in 1301), but admits that its origin is uncertain and that 'no further related word is known outside English'. It suggests that the word is identical in form to the same word meaning 'magpie' which 'is held by many to have been in some way derived from or connected with that word'. The suggested connection is that a pie has contents of a 'miscellaneous nature', similar to the magpie's colouring or to the odds and ends picked up and used by the bird to adorn its nest. As support for the argument the *Oxford English Dictionary* notes the similarity between the words haggis (the Scots dish with contents of a miscellaneous nature) and an old French word for magpie (*agace or agasse*).

The only other possible origin mooted by the *Oxford English Dictionary* is that the word 'pie'—meaning a baked pastry—may be connected with the same word used in an agricultural context for 'a collection of things made into a heap', such as potatoes or other produce covered with earth and straw for storage. But perhaps there is another clue buried in the *Oxford English Dictionary*. The Gaelic word *pige* (or one of its variants) may be related to the English dialect word *piggin*, which can mean a wooden pail, or an earthen pot such as might be used for cooking. If a non-linguist may be permitted an idea, could there be a connection here? The early history of the pie, which will be explored in chapter One, would suggest that this is possible, and it is surely no less tenuous a connection than that with a bird?

A French linguistic connection cannot be dismissed completely either. The English language was never the same again after the Normans invaded in 1066, and a whole lot of pie-words are similar enough in French and English to suggest a similar origin—think tart and *tourte*, for example. We will come across a few others in later chapters, but the most intriguing are the words paste, pastry and pasta, which are all variations on a theme of flour and water, and have a common origin. The French word pâté comes from the same root, which is more obvious when you discover that the circumflex over the 'a' replaces a missing 's', indicating that a pâté used to have a pastry crust. Modern pâté is pastry-less, so the original form now has to be qualified as a pâté *en croûte* (*croûte* meaning 'crust'). The most fascinating connection is that 'pasty' also appears to come from this same root, raising the question of why this specific type of pastry-wrapped food got a name with a totally different origin to that of the ordinary 'pie'.

Chapter 1   A Brief History of Pie

> Pyes were but indigested lumps of dough,
>
> Till time and just expence improv'd them so.
>
> *Of Apple-Pyes; A po em*, by Mr Welsted (1750)

Once upon a time, *everything* baked in an oven that was not bread was 'pie'. *Everything*. Let me explain.

It seems likely that the earliest ovens were actually kilns, used to fire clay objects such as figurines and pots. Bread, meanwhile, was baked in flat cakes on hot hearthstones. Someone, somewhere, perhaps, noted the similarity between clay and dough and, in an act born either of inspired impulse or thoughtful experimentation, tossed the dough into the kiln instead. The bread oven was born. Meat, meanwhile, was cooked by direct exposure to the fire, on a spit or in the coals. The problem with cooking meat this way is that even if it does not burn, the valuable and tasty juices drip away and the meat dries and shrinks. Other cooks at other times got around this problem by wrapping the meat up to protect it—in leaves, for example. Or clay. Clay that, to another cook in perhaps another time and place, felt just like dough. This last inspired step created the primitive meat pie—something medieval cooks called a 'bake-mete'.

The thick crust of this early pie acted like a baking dish. For hundreds of years it was the only form of baking container—meaning *everything* was pie. The crust also, as it turned out, performed two other useful functions: it acted as a carrying and storage container (before lunch boxes) and, by virtue of excluding air, as a method of preservation (before canning and refrigeration). These early piecrusts were called 'coffins', which sounds vaguely sinister today, as if our ancestors were already implying that the contents of a pie could be doubtful. In actual fact the word originally meant a basket or box (think of a *coffret* of perfume), and was used in relation to pastry 'caskets' before it came to refer to the funeral variety.

But what was this crust actually like to eat? We know that it was often made several inches thick to withstand many hours of baking, so it would have been very hard. It is usually referred to in modern texts as inedible, and *not meant to be eaten.* As a blanket statement I find this hard to believe. Certainly to our delicate modern sensibilities, it sounds unpalatable—but we are talking about harder times, when growing and harvesting grain and producing flour—even very coarse flour—was an incredibly labour-intensive process. It is surely not likely that such a hard-won resource was simply discarded after the contents were eaten, even in the great houses? The crust may not have been intended for lords and ladies, but the well-to-do were obliged to feed their servants and were also expected to feed the local poor. Would not this largesse of sauce-soaked crust be distributed to the scullery boys and the hungry clamouring at the gate?

There are suggestions in early manuscripts that the crusts were, at least occasionally, reused. One fifteenth-century manuscript for 'Fresh Lamprey Bakyn' (a lamprey pie) suggests that, after the lamprey is eaten, the remaining juice may be boiled up with wine, sugar and spices and then poured back into the coffin, which has been refilled with layers of 'white brede', creating a new dish called 'soppys in galentyn'. Sometimes the baked dough was recycled as a thickening agent, such as in one seventeenth-century recipe for a 'Spanish Olio' which suggests that 'some Crusts of Bread, or Venison Pye-Crust' be added before the stew is boiled 'in all five or six hours gently'. In other words, it was used as we would use a *roux*.

Thankfully, the problem of what to do with excruciatingly hard leftover baked dough eventually went away, as we will see.

The Invention of Pastry

After a few millennia of inspiration, the primitive clay oven gave rise to the gleaming modern steel version. A high-tech oven alone does not, however, turn a bake-mete into a pie as we know it. One more important development was necessary. Pastry had to be invented.

Dough becomes pastry when fat is added. Not just any

dough, not just any fat, and not just by any random method of mixing. Exactly when and where and how this happened are mysteries that we will never completely solve as it all took place long before our earliest surviving cookbooks were written. We can make some educated guesses, however, and some of our clues come from considering the crucial difference between 'dough' and 'pastry'. For this we need to make a brief foray into food chemistry.

The key is gluten. Gluten is a protein with long, elastic molecules which simultaneously enable the dough to be made stronger (by providing structure) and lighter (by enabling the trapping of air bubbles). A lot of gluten means a firm structure, which is ideal for bread, but bad for pastry. Too little gluten means no structure and no air-trapping, so flat bread and tough pastry. The task of the pastry-cook is to get just the right amount of gluten to make the pastry light and crumbly and flaky. How is this done?

Wheat is the only grain with a significant amount of gluten, so we have our first clue to the origins of pastry. Superb pastry could only have developed where wheat was grown: rye, barley and oats do not make good pastry, nor do rice or maize or potato starch. The gluten content of the final wheat dough can be manipulated by the cook in a number of ways depending on the ultimate goal, whether it be sturdy bread or flaky pastry. These tricks of the pastry-cook's trade give us further clues to the development of pastry itself.

The first trick is to use exactly the right amount of water, for it is water that activates the gluten in flour.

Fat is the second trick, and it helps the texture of pastry in a number of ways. Fat coats little packets of flour, waterproofing them and limiting the amount of water that gets in (less water, so less gluten), and it keeps the gluten strands 'short' and slippery. Little smears and gobbets of fat also physically separate the mini layers of dough, so that they form individual flakes or crumbs, not a solid mass. It goes without saying, then, that the proportion of fat to flour used is crucial in controlling the final texture of the pastry, but it is not the only factor.

Like flours, not all fats are created equal. Oil is fat that is liquid at room temperature—there is no other essential difference—but good pastry cannot be made with oil. Flour simply absorbs the oil and the resulting dough is mealy, not tender and flaky. The ideal fat for pastry-making is one with a high melting point because the longer it takes for the fat to melt, the longer it keeps the little parcels of dough separate, generating little packets of steam to puff and lighten the dough. Pig fat (lard) has a high melting point and very little water content, so is ideal on both counts. Butter melts at body temperature so does not give quite such a good texture, but the trade-off is a richer flavour. So now we have our second clue to the whereabouts of pastry development: truly delicious, light, 'short' pastry could only develop where there was a good supply of solid fat, which in essence means where there are pigs or cows.

Pastry-making, as every amateur baker fears, is as much about technique as ingredients. The rationale behind the well-known advice to keep the hands, implements and kitchen cool while making pastry, to use minimal water and to handle it lightly is obvious, now that we understand the process. Cool handling lengthens the time that the fat in the dough stays solid; using the minimum amount of water reduces the gluten content and also allows the dough to be crisper; minimal handling also reduces the gluten, so we do not knead pastry dough as we do bread.

As far as a time-frame for these developments goes, we can probably reasonably deduce that the pie began its life some time before the fourteenth century in those areas of Europe where wheat was grown and pigs and cattle reared. There does not seem to be any doubt that a huge variety of pies was made in northern and central Europe throughout the Middle Ages, and this was also partly because the forests of northern Europe provided an abundance of fuel (and fodder for pigs) long after those of the south had been depleted. Why is this important? Because the grand bake-metes and pies (which might contain a whole haunch of venison) took a long time to bake and required a lot of fuel.

Size is not the only thing that matters in pies of course. Another branch of the pastry art resulted in the elegant tarts and small pastries—particularly the sweet ones—that most of us find irresistible. It seems likely that the basic idea of small sweet pastries came to Europe from the Arab world during the expansion of the Muslim empire in the seventh century, but it is probable that it was the northern Italians of the Renaissance who refined and developed pastry into the many glorious incarnations we appreciate today. All of the resources were in place, and the context was right. Wheat from northern Italy is 'soft'—that is, it is already low in gluten so is ideal for pastry-making. Butter was the fat of choice for cooking in northern Italy (and a sign of wealth), compared with the oil of the south of the country, and there is no doubt that butter makes the finest pastry for sweet pies and tarts. The philosophical environment of the Renaissance encouraged the development of all the arts, and the dynastic families of Italy had no shortage of money for indulging their lust for the good things of life. Brilliant and innovative cooks were coveted and encouraged, and there was a grand blossoming of culinary ideas and techniques which gradually spread to France and the rest of Europe.

The situation in Britain was different. In Britain, butter was food for the poor. The wealthy in Britain preferred lard, maybe because the animal had to be killed to obtain the fat, thus its perceived value was higher. Lard makes superb huge 'raised' or 'standing' pies full of meat, which flourished to become one of the jewels in England's culinary crown.

The argument about whether or not medieval pastry was meant to be eaten is a relative one. What we really want to know is how much of it was designed to be eaten as an intrinsic and desirable part of the dish. There are suggestions in the earliest cookbooks that at least some of the time this was certainly the case. Why else would a cookbook writer specify that the pastry be made thin and 'tender as ye maye'? This instruction appears in the first-known written recipe for pastry, in a little book called *A Propre New Booke of Cokery*, published in London in 1545:

To make short paest for tarte

Take fyne floure and a cursey of fayre water and a dyshe of swete butter and a lyttel saffron, and the yolckes of two egges and make it thynne and tender as ye maye.

The instructions in early cookbooks are, for a food historian, frustratingly lacking in detail as they were based on much assumed knowledge, and were intended as *aides-memoires* for experienced cooks. It is clear, however, that by the sixteenth century short, puff and perhaps choux pastries were already established, and there are tantalizing clues that they may have already been around for a couple of hundred years. The earliest-known cookbook from England, *The Forme Of Cury*, was compiled in about 1390 by the master-cooks of King Richard II, and it contains instructions for 'payn puff', which sounds suspiciously like puff pastry.

There is one particular type of pastry that has played an important role in the history of the pie and that breaks a few of the method rules. It spans the bridge between bread dough and regular pastry, and can still be found in the traditional English pork pie such as that made famous in Melton Mowbray in Leicestershire. It is what we refer to now as 'hot water pastry'. Its great advantage over bread dough and other types of pastry is that it can be sculpted like clay, allowing it to be 'raised' to form free-standing crusts. A whole lot more possibilities opened up when raised crusts were developed, as they allowed 'wet' fillings such as stews, fruits and custards to be cooked in them.

Before baking dishes, then, dough in one form or another was used as the 'container'—so by definition everything cooked in an oven was 'pie'. The legacy of this persists in a number of food words that don't at first glance seem to have anything to do with pie. 'Custard' comes from *croustade* or crust, which the *Oxford English Dictionary* defines as 'formerly, a kind of open pie containing pieces of meat or fruit covered with a preparation of broth or milk, thickened with eggs, sweetened, and seasoned with spices, etc.'. A 'dariole' is now a small mould for baking soft puddings or creams, but once it meant a small pie 'filled with flesh, hearbes, and spices, mingled and minced together'. Even a 'rissole' was once 'a sort of minced pie' that was usually fried (from the French *rissoler*, to fry brown).

These adapted pie-words are the lucky ones—many more have completely disappeared. When was the last time you saw on a menu a *chewet* (a small round pie of finely chopped meat or fish, with spices and fruit, 'made taller than a marrow pie'), a *dowlet* (a small pie of particularly dainty little tidbits), a *herbelade or hebolace* (a pie with a pork mince and herb mixture), a *talemouse* (a sort of cheesecake, sometimes triangular in shape) or a *vaunt* (a type of fruit pie)? These words (and more) were once everyday words in a baker's vocabulary. The only conclusion it is possible to draw is that the loss of so many pie-words reflects the loss of the pies themselves.

Is the Pie Dead?

The importance of the pie—once the 'meat and potatoes' of the English—began to slip with the increased cultivation of the actual potato in the nineteenth century. As the nineteenth became the twentieth century, social changes pushed the pie further into decline. The 'great pies' had their last glorious days in the English manor houses of the Edwardian era, before the domestic classes left to fight the First World War. The ordinary family dinner pie hung onto its place a little longer—it was alive and well during the Second World War—until the housewife increasingly became a working wife with less time for cooking. Since then, our sense of being time-poor has escalated enormously. Naturally, we gave up the most time-consuming and intimidating cooking first. Simultaneously, we have been subjected to the overwhelming propaganda of the nutrition police. Whichever way you look at it, the homemade pie has been under siege for a century at least, and surely its survival is endangered.

The strangely marvellous thing is that, we refuse to relinquish the pie. We cling to the *idea* of it with some fervour, in spite of its fading reality on our tables. Why is it so? What is it *about* pies?

Chapter 2   The Universal Appeal of Pie

A boy doesn't have to go to war to be a hero; he can say he doesn't like
pie when he sees there isn't enough to go around.

Edgar Watson Howe

There was no doubt in the minds of nineteenth-century cooks and cookbook writers that there was *something* about pie—a difficult to grasp something that made it universally esteemed in a way that cake or stew or soup was not. A few quotations will suffice to demonstrate.

In 1806, Mrs Rundell began her 'Observations on Savoury Pies' in her bestseller, *A New System of Domestic Cookery*, with the confident statement: 'There are few articles of cookery more generally liked than relishing pies, if properly made.' The celebrated chef Alexis Soyer noted the everyday importance of pies in the Victorian era in his *Shilling Cookery for the People* (1860):

> From childhood we eat pies—from girlhood to boyhood we eat pies—from middle age to old age we eat pies—in fact, pies in England may be considered as one of

our best companions *du voyage* through life. It is we who leave them behind, not they who leave us; for our children and grandchildren will be as fond of pies as we have been; therefore it is needful that we should learn how to make them, and make them well! Believe me, I am not jesting, but if all the spoilt pies made in London on one single Sunday were to be exhibited in a row beside a railway line, it would take above an hour by special train to pass in review these culinary victims.

The pie was placed on an even higher pedestal by the social journalist, Charles Manby Smith in his book *Curiosities of London Life* (1853), where he declared it 'a great human discovery which has universal estimation among all civilized eaters'.

There will always be naysayers of course, but the few dyspeptic curmudgeons and ascetics who disagree on the appeal of the pie only serve to prove the rule. Ambrose Bierce in his *Devil's Dictionary* was perhaps merely demonstrating his need for medical advice when he defined the pie as 'an advance agent of the reaper whose name is Indigestion'. The nineteenth-century nutrition guru Sylvester Graham managed to convince a significant percentage of the American public that the answer to all of society's problems and the guarantee of heavenly reward was a strict regime of cold baths, bland food and sexual restraint, and that cholera was caused by chicken pies. Had he lived a few decades longer and seen the evidence that cholera was due to contaminated water, poor Graham may have allowed himself an occasional pie to compensate for all that sexual restraint.

The Usefulness of Pie

The original pie served three very useful functions in acting as a baking, carrying and preserving container. We now have more efficient ways to perform these functions, but the pie is still an extraordinarily pragmatic food for both cooks and consumers. For sheer versatility the pie is impossible to beat. It is useful in ways that bread (very useful) and soup (somewhat useful) are not, precisely because of this versatility. Pies can be eaten hot or cold, at every course of every meal from breakfast to supper, not forgetting at picnics and while travelling, and they are especially suited to meal-in-the-hand events. They can be economical or extravagant, an everyday meal or a special feast, a simple food or a powerful symbol. The fillings can be as varied as circumstances, imagination and conscience will allow, and the crust is a superb opportunity to show off the artistic as well as the culinary skill of the cook.

An outstanding historical feature of the pie is that it is a self-contained meal which can be eaten in the hand, without need of cutlery, crockery or napery. This ultra-convenient aspect of the pie might be one reason for its success in England. Sheila Hutchins, in her book *English Recipes* (1967), says: 'The sporting English aristocrats with their passionate interest in cock-fighting or cricket had early developed a system for eating well without interfering with whatever might be on hand.' Hutchins illustrates this with the story of the Earl of Sandwich, reluctant to leave the gaming tables one night in 1762 and calling for a piece of beef between two slices of bread, immortalizing his name in the process. It is an amusing myth which I do not understand. The era and the country were famous for pies. Why would he not have called for a bit of pie left over from dinner?

The reheatability of pies can be both an advantage (usually to the cook) and a disadvantage (usually to the consumer). Chaucer knew this well in the fourteenth century, and illustrated it beautifully in his *Canterbury Tales*. The Cook is challenged to tell his own story to make up for serving reheated pies (from which he had already siphoned off some of the gravy) to the pilgrims:

> Now telle on, Roger; looke that it be good,
> For many a pastee hastow laten blood,
> And many a Jakke of Dovere hastow soold
> That hath been twies hoot and twies coold

A 'Jack of Dover' was a bottle of wine made up of collected dregs from other bottles, recorked and sold as new. Chaucer uses this as a metaphor for the repeatedly reheated pies, making the idea even more distasteful by describing the Cook as a dirty man with suppurating sores on his legs.

When Soyer said of pies that they are 'one of our best

companions *du voyage* through life', he was referring to consumers, but he might just as well have been referring to his professional colleagues, for pies have always been enormously useful to caterers and cooks, particularly at events where a large number must be fed efficiently. In modern times this is usually at sporting events such as football games, but the original experts in mass catering were the military. On 11 July 1891 the visiting German emperor reviewed the troops at Wimbledon. The Post Office Volunteers had Quartermaster Dickson in charge, and he invented an efficient, spectacularly simple system which impressed the journalist who described it:

> Each man as he passed took without halting a tankard in one hand and received a pork pie in the other, then passing on to enjoy his luncheon a hundred yards off. The result was that about 800 men were served in exactly seven and a half minutes.

Quartermaster Dickson was way ahead of his time. Eight hundred customers in seven and a half minutes would be an amazing achievement for the fastest modern fast-food outlet.

Biology Makes Us Do It

There is no escaping a biological fact that is the despair of dieters everywhere. For survival reasons we are genetically programmed to crave nutrient-dense foods (such as pies), eat as much as possible of them while they are available and convert them efficiently into body fat in readiness for the next famine. Primitive humans did not need to develop a craving for fibre and antioxidants: life was already nasty, brutish and far too short for these to add any value. So, how did early *Homo sapiens* determine what foods are nutrient-dense without the help of nutritionists and dieticians?

We are attracted first, by smell. Our noses detect the volatile components in our chemical environment, and these are increased by heating. We should expect then that the smell of cooked high-fat, high-protein foods would be the most appealing to us as a species, and it appears that this is indeed the case. At a simple biological level, before dieticians were invented, our noses decided what was best to eat: if it smelled good, it was almost certainly good to eat. When VAT (British goods and services tax) legislation was first being formulated in the UK, bakers successfully argued that hot meat pies should be VAT-free as the major reason for having them hot on the premises was to create an enticing smell, rather than to provide a special service—and even the most uninspired pie smells enticing while it is hot.

For a long time in the West it was believed that our taste buds could only determine four basic flavours—sweet, salty, sour and bitter. Most scientists now accept what the Japanese have known for hundreds of years, that there is a fifth taste called *umami*. This is a concept of 'delicious flavour' or 'savouriness' such as that we taste in aged cheeses, roasted meat and Asian fish sauce. The taste is conveyed primarily by glutamates, which are amino acids, which in turn are the building blocks of protein. It makes sense, biologically, that we would have built-in protein detectors. Sweet foods are a source of calories, salt foods help us keep our chemistry in balance, sour and bitter tastes make us wary as in nature they often signify a poison. *Umami* wakes up our protein sensors. Mother's milk has *umami* (ten times the glutamates of cows' milk). Meat pies have *umami* too.

There is also a third process going on when we eat: the tactile appreciation of the food, the sense of its 'body' or richness—that sensation we call 'mouthfeel'. If our enjoyment of food were simply about taste and smell (and perhaps, now that we are civilized, appearance) we could purée a pie, put it in a glass and decorate it with a paper umbrella and still enjoy it. A pie is only as good as its pastry, and one of the delights of a good pie is the contrast in texture between the crisp pastry and the filling—whatever it might be. In a perfect pie, each component is independently perfect—the mouthfeel of the pastry (buttery, flaky, crumbly) and the mouthfeel of the filling (rich, unctuous, tender, sticky, crunchy etc.); and the whole is more than the sum of its parts.

So is the pie not the perfect food for the senses? Does any other single dish have such potential?

Beyond Calories

It seems that the *liking* of pies has a sound biological base, but nutrient density alone is not sufficient for modern humans, or we would eat whale blubber with the same relish that we eat pie. We are social animals, and we don't usually find and eat food alone, so we associate it at an emotional level with people, events and circumstances. Eventually a food becomes embedded with meaning, allowing anthropologists to ask questions like: 'Do pies *mean* anything?'

Smell can be a powerful trigger of emotion because humans make associations between smells and the social and emotional context in which they are perceived. Some intriguing work has been done on 'olfactory evoked recall' at the Smell and Taste Treatment and Research Foundation in Chicago. Of course, the subjects were all from the Chicago area, and likely had European ancestors (as does pie), so the results may not apply to other demographics. Nevertheless, the results are interesting. The smells giving rise to the most nostalgia were from baked foods (bread, cakes etc.), closely followed by those from cooked meat dishes such as bacon or meatballs and spaghetti. It sounds to me that if there had been a separate meat pie category it would have combined the associations of both and won hands down.

This is nothing new to many non-scientists of course. Estate agents have long known it, which is why they advise vendors to have a pot of freshly brewed coffee in the house to make it smell like a home when potential buyers visit. Proust explained it better than anyone before or since when he described the flood of memories triggered by the scent of a buttery little cake dipped in fragrant lime-blossom tea.

Craig Claiborne said: 'I have learned that nothing can equal the universal appeal of the food of one's childhood and early youth.' It is the food that looks backwards through our shared family memories. It is comfort food, the food inextricably linked in our cultural consciousness with motherhood and nationhood. Even though pies are no longer a daily item on our dinner tables, they still figure large in many of our memories: pies mean Thanksgiving and Christmas and picnics and silly old Aunt Mabel and going to the football with Dad. The pie-cook and the pie-consumer are both lucky if the smell of the pie 'sells' not only its desirability as biological fuel but also the remembrance of pies past.

There is one other meaning of pie, particularly the homemade one, which sums up its universal esteem. In the words of Margaret Fulton, 'A pie is invariably acclaimed as a treat and a sign of a caring cook.' A cook who has gone to extra trouble, who loves you enough not just to toss the stew onto a plate with a lump of bread, but to craft for it its own little pastry gift-box. Is this the crux of it?

Chapter 3   Pies by Design

The fine arts are five in number, namely: painting, sculpture, poetry,
music, and architecture, the principal branch of the latter being pastry.

Antonin Carême (Marie-Antoine Carême)

It is ironic that Antonin Carême (1784–1833), a man
revered as 'the chef of kings and king of chefs', had he had a
choice, would probably have studied architecture. His culinary
career came about by default when, as a child abandoned by
destitute parents, he found work as a lowly kitchen boy. He
spent the rest of his working life in the kitchen, but at heart
he remained an architect, expressing his passion through the
design and construction of the incredibly ornate *pièces montées*
which were his signature.

Maybe Carême had a point. A pie, like a building,
requires construction after all. Also like a building, one
size and style of pie does not fit all fillings and occasions.
Architects of the early twentieth century may have thought
they invented the idea of 'form follows function', but pastry-
cooks had been quietly applying the principle for hundreds
of years. A pie containing a whole side of venison and

intended to preserve the meat for a prolonged period of time is a design-world away from a delicate almond-milk custard acceptable for Lent, and a whole design-universe away from one containing the entirely non-culinary treat of a very alive, scantily clad young woman. Other circumstances dictate other designs: a pie might need to be robust enough to send on a long sea voyage, elegant enough to serve as a gift, clever enough to entertain, or awe-inspiring enough to deliver a message of power or propaganda.

Luckily for the housewife, as the seventeenth century progressed an increasing number of cookbooks were written for the non-professional, meaning that less knowledge was assumed. Luckily for us, they therefore give us more insight into the pastry and pie-making business. Gervase Markham in his *English Hus-wife* (1615) went to some pains to clarify the different sorts of pastry that were appropriate for different pies.

> Our English Hus-wife must be skilfull I pastery, and know how and in what manner to bake all sorts of meate, and what paste is fit for every meate, and how to handle and compound such pastes as for example, red Deere venison, wilde Boare, Gammons of Bacon, Swans, Elkes, Porpus, and such like standing dishes, which must be kept long, wold be bak't in a moyst, thicke rough course, and long lasting crust, and therefore of all other, your Rye-paste is best for that purpose: your Turkie, Capon, Pheasant, Partridge, Veale, Peacocks, Lambe, and all sorts

of water-fowle, which are to come to the Table more then once (yet not many daies) would be bak't in a good white crust, somewhat thicke; therefore your Wheat is fit for them: your Chickens, Calvesfeet, Olives, Potatoes, Quinces, Fallow Deere and such like, which are most commonly eaten hot, would be in the finest, shortest and thinnest crust; and therefore your fine wheate flower which is a little baked in the oven before it be kneaded is the best for that purpose.

Markham makes a point about pies 'which must be kept long' because this was one of their most important requirements in his day. Before refrigeration and canning were invented the only ways that meat could be preserved were by drying, smoking, salting—or encasing in a 'thicke rough course crust'. The pastry for pies 'to be kept long' was usually of rye flour, several inches thick, baked until very hard and 'not proposed for eating, but to keep the Inside properly'. When the pies were taken out of the oven, melted fat was poured in through a hole in the lid to exclude air, thus preserving the contents. Once the pie was cut this airtight seal was broken, leaving the contents prone to rapid spoilage—which perhaps gave rise to the old superstition that it is unlucky to take just one slice from a pie.

Some of the instructions in old cookbooks that don't make much sense to us today were perhaps intended to maximize the keeping potential of the pie. One fifteenth-century manuscript has an odd warning against getting saffron 'nygh the brinkes' (near the edges) of your pie, 'for then hit will never close'. As saffron was a common ingredient in both pastry and pie fillings at the time, this instruction is inexplicable. The instructions for joining pie-lid to base in the sixteenth-century German cookbook of Sabina Welserin start in a recognizable way for us with 'join it together well with the fingers', but then she advises:

> Leave a small hole. And see that it is pressed together well, so that it does not come open. Blow in the small hole which you have left, then the cover will lift itself up. Then quickly press the hole closed.

Was this just to make a nice domed top, or was it also to lift the dough off the wet contents? It was essential for the coffin crust to be kept very dry or it would lose its preserving power. In William Salmon's *Family-Dictionary*, or, *Houshold Companion* (1695) he gives a recipe for a boar pie sealed with butter that 'will, if it be not set in a very moist place, keep a whole Year'. Keeping a meat pie for a whole year without refrigeration is a terrifying thought today, but it was such a common practice that we have to assume that most of the time consumers survived the experience.

A thick dense crust also formed a robust container that enabled pies to be sent long distances over land and sea. There was a regular traffic in grand 'Yorkshire Christmas Pies' to London by rail in the nineteenth century, but this was nothing compared to some of the efforts made by devoted mothers and wives over the centuries in getting pies to their sons and husbands away at university or at war.

One worried mother was Lady Brilliana Harley of Brampton Bryan in Herefordshire. She regularly sent pies by carrier to her son Edward at university in Oxford in 1638–1639, a trip that would have taken several days by carrier. Ned survived his university days so it is to be assumed that her pies were robust and well sealed. On 10 May it was a kid pie:

> I have made a pye to send to you; it is a kide pye. I beleeve you have not that meate ordinaryly at Oxford; one halfe is seasned with one kind of seasening, and the other with another.

Once the postal service was established, pies were sent by mail to destinations far and wide. The Post Office gave excellent advice in 1884 on how to do this at Christmas:

> It is desirable that any Christmas presents—such, for instance, as contain holly, mistletoe, or other decorations, poultry, game, puddings, mince pies, or any other pastry, confectionary, apples, toys, fancy articles &c,—intended for transmission by parcel post should be carefully packed by the senders so as to preserve them from injury.

The very reliable Post Office even managed to avoid pilfering of the goodies en route. One English soldier in

South Africa during the Boer War was moved to write to *The Times*:

> I should like to say a word in praise of the postal and parcel arrangements out at the front . . . I am confident I have not missed a single parcel ... in spite of the fact that I am continually imploring my people not to label parcels chocolate, game pie, chicken and tongue &c.

The Art of Pies

The architects who declared that 'form follows function' also declared that all ornamentation is bad. It is abundantly clear that pastry-cooks have never had any inclination to follow this philosophy. Pastry is an eminently artistic medium and generations of practitioners of the 'curious art of pastery' (curious as in its old meaning of 'ingenious, skilful, clever') have enthusiastically embraced its potential.

The technological advances of the Industrial Revolution enabled the manufacture of incredibly elaborate pie moulds, but well before that time pastry-cooks took advantage of the mouldability of hot-water crust to create pies 'made craftily in the lyknes of a byrde's bodye', or a castle, or a fish, or any other shape that that the baker's inner artist desired. Pies in the form of castles had an enduring popularity (perhaps all early pastry-cooks were frustrated architects?) There is a recipe in the *Forme of Cury* for a 'chastlete', which is a pastry castle with four crenellated towers around a central courtyard, each with a different filling and colour. At the end of the eighteenth century, Parson James Woodforde noted in his diary a meal which included 'beef-stake tarts in turretts of paste'.

Sometimes the lid of a large coffin was removed before serving, and replaced with a separately baked lid decorated to look like a heraldic shield, for example, or the lid was left off and the surface of the filling 'made as gay as you please' with herbs or flowers, or shapes cut from pastry or coloured jelly. One recipe from 1658 for a steak pie instructs that the lid be removed and fried sage leaves be stuck upright in the walls before serving, which must have made it look like a little indoor garden.

The seventeenth-century baker could call on paper templates 'cut into divers proportions, as Beasts, Birds, Arms, Knots, Flowers, and such like' to make pastry shapes which were often baked in advance, ready to decorate the tops of pies before serving. There were other parallels with needlework too: 'stump pies' were so called because they were embellished with such intricate pastry knots, flowers, heraldic symbols and so on that they resembled the three-dimensional form of embroidery popular in the Jacobean period known as 'stump work'.

Pastry artists of previous times also worked with colour: pie-lids might be frosted (with sugar mixed with rosewater), 'endored' ('gilded' with saffron, egg yolk or real gold), or painted with various colouring agents. Recipes for colouring agents appear in many books from the medieval period onwards, and some of them are very alarming. *The Widowes Treasure* (1586) gives a recipe for *An Emeraulde Greene* containing *verdegres* (copper acetate), *litarge* (lead oxide) and quicksilver (metallic mercury) mixed with 'the pisse of a young childe', and another for a gold colour containing saffron, *orpiment* (arsenic trisulphide) and the gall (bile) of a hare or a pike pounded together, placed in a vial, and buried in a dunghill for five days. It is perhaps better after all that some techniques have died out.

There was also an art to serving a pie properly in the days when they were often very large and were intended to feed many people. The elegant and correct carving of different joints of meat was an essential skill and an honoured role for a gentleman, who was also expected to know the correct terminology. He would, for example, 'dysmembre that heron, traunche that sturgyon, lyste that swanne, disfygure that pecocke' and 'border that pie'. He would also know that a pie should be opened at the top if it was to be served hot, and 'midways' if cold.

Naturally, if one is not to seem boorish, one should know how to eat a pie correctly. Thankfully, etiquette manuals have been around for centuries to show how this should be done. A book of manners published in 1609 and addressed to children reminded them that 'If a peece of pie or of tart, be offered thee, receive that on thy plate or trencher, and not with thy fingers.' In 1853 ladies were advised:

It is an affectation of ultra-fashion to eat pie with a fork, and has a very awkward and inconvenient look. Cut it up first with your knife and fork both; then proceed to eat it with the fork in your right hand . . . At a public table, a lady should never volunteer to dress salad for others of the company. Neither should she cut up a pie, and help it round. These things ought only to be done by a gentleman, or a servant.

The Social Life of Pies

In modern society, where anyone in theory can make money, it is difficult to appreciate that once upon a time wealth was tied absolutely to social class, and therefore social class determined what you ate, even to the extent of determining the type of pastry making up your pie. Farming and house- hold manuals of the seventeenth and eighteenth centuries clearly instructed that the piecrust for the master's family be made from the finest wheat flour, whereas for the servants' piecrust the second milling of wheat or barley was to be used, or maslin (a mix of wheat and rye) or rye.

Game pies and many fish pies by definition could only be enjoyed by those who had the land and the hunting rights. The poor and the working class of Victorian England got mutton pies (if they were lucky), with meat from sheep past their prime, or beef pies from old dairy or draft animals. An English newspaper article of 1857 about the famous Derby horse race meeting summed up the class-food divide nicely:

> The Derby is worth seeing. I do not know where
> England altogether, is so well represented. It is there
> in samples—the highest aristocracy and the lowest
> democracy . . . The Bishops drive out in their coaches
> with hampers of game pies and champagne; and the
> costermonger, loaded with bread, cheese and beer, drives
> out with his barrow and donkey.

In England, pigeon pies were high-status pies for high-status people because only wealthy landowners could maintain pigeon cotes to supply fresh meat over winter. In America the situation was initially the reverse. Early in the nineteenth century vast flocks of passenger pigeons, several miles across, would literally darken the sky. The birds were freely and extraordinarily easily available for anyone with a gun and a craving for pigeon pie dinner. So successful were the hunters that the passenger pigeon became extinct by 1914, but in the small hiatus before their end, when they were a rarity, they briefly became expensive restaurant food for the wealthy.

The dynasty of pies had its own hierarchical order on the eighteenth-century dining table. At that time meals were served in the style that came to be called *à la française* (as distinct from the style we use today, service *à la russe*, in which individual dishes are served sequentially to guests). Two or more courses would each consist of a variety of dishes set out simultaneously on the table with geometric precision and an eye for symmetry. Pies were very important in this style of service, as they had impressive visual 'presence' on the table. Grand pies often formed the centrepiece, with smaller pies at the sides and corners (some were actually called 'corner pies'). The range of possible pie fillings at this time in history was truly amazing.

Chapter 4   Filling Stuff

Good apple pies are a considerable part of our domestic happiness.

Jane Austen

The *Oxford English Dictionary* gives one definition of 'stuff' as 'materials for filling a pie'. It does not attempt to list the various materials—nor could it, as almost everything imaginable from the sublime to the sinister has at some time or other found its way into a pie.

The original pie, as we have seen, contained single large pieces of meat. By the fourteenth century the range of fillings had broadened to include fruit, delicate custards and finely minced and spiced meats. One thing that strikes us today when we look at recipes of this era is that many of the 'savoury' dishes contain sugar. The distinction between sweet and savoury food is a relatively modern one; medieval cooks knew no such distinction. Sugar was an expensive imported ingredient and was used in the same way as a spice. It became cheaper with the opening up of sugar refineries in Britain in the mid-sixteenth century, and cheaper still with the development of the East India trade in the seventeenth century.

By the mid-seventeenth century, sugar was not used sparingly as a spice, but in quantities that clearly imply a taste for sweetness. The famous bestselling seventeenth-century cookbook *The Accomplish't Cook* by Robert May included this fascinating recipe for herring pie:

To make minced Herring Pies

Take salt herrings being watered, crush them between your hands, and you shall loose the fish from the skin, take off the skin whole, and lay them in a dish; then have a pound of almond paste ready, mince the herrings, and stamp them with the almond paste, two of the milts or rows, five or six dates, some grated manchet, sugar, sack, rose-water and saffron, make the composition somewhat stiff, and fill the skins, put butter in the bottom of your pie, lay on the herring, and on them dates and gooseberries, currants, barberries, and butter, close it up and bake it, being baked with butter, verjuyce, and sugar.

By the eighteenth century the distinction between savoury and sweet dishes was becoming clearer, and many cookbooks of the time gave two versions of meat pies to cater for alternative taste preferences. This recipe from Hannah Glasse's *The Art of Cookery, Made Plain and Easy* (1747) was followed by the savoury alternative.

To make a very fine Sweet Lamb or Veal Pye

Season your Lamb with Salt, Pepper, Cloves, Mace

and Nutmeg, all beat fine, to your Palate. Cut your Lamb, or Veal, into little Pieces, make a good Puff-paste Crust, lay it into your Dish, then lay in your Meat, strew on it some stoned Raisins and Currants clean washed, and some Sugar; then lay on it some Forced meat Balls made sweet, and in the Summer some Artichoke Bottoms boiled, and scalded Grapes in the Winter. Boil Spanish Potatoes cut in Pieces, candied Citron, candied Orange and Lemon-peel, and three or four large Blades of Mace; put Butter on the Top, close up your Pye, and bake it. Have ready against it comes out of the Oven a Caudle made thus: Take a Pint of White Wine, and mix in the Yolks of three Eggs, stir it well together over the Fire, one Way, all the Time till it is thick; then take it off, stir in Sugar enough to sweeten it, and squeeze in the Juice of a Lemon; pour it hot into your Pye, and close it up again. Send it hot to Table.

Minced meat, spices, sugar, dried fruit—there are medieval echoes to be heard in this recipe, and they can still be heard in our Christmas mince pies. The echo persists more strongly in, of all places, a little town called Pézenas in the Languedoc region of France. A local speciality is its *petits pâtés de Pézenas*—tiny, bite-sized plant-pot-shaped pies with a sweet lamb filling. The legendary Lord Clive of India stayed in Pézenas for several months in 1786 and, in spite of poor health, led a busy social life. Locals were intrigued by the pies served at his home, and prevailed upon his cooks for the recipe. Once they became a local speciality, the pies became immutable to change and endured like culinary time- capsules, giving us a glimpse into a previous era. A similar thing happened in Nova Scotia to the Cape Breton pork pie, which contains no pork—but has dates where pork presumably used to be.

Fish Pies

Fish is a perfectly good alternative to meat as a pie filling—if you have a choice, that is. For many periods in history, meat-less days were mandated by the Church (and often backed by statute law) and at some periods these added up to almost half the days on the calendar. Fish pies took on a great significance on these days, particularly for grand occasions.

There were a number of reasons for decreeing abstention from meat. In ancient times meat was thought to inflame the passions (thereby distracting the mind from higher thoughts) whereas fish (or rather, creatures that lived in the water, which included whales and 'porpuses') were seen as cooling. It was also believed that the characteristics or habits of everything in the natural world would be transmitted to the eater, so the fact that fish did not have an obvious sex life added to its suitability for days of religious observance. In later periods the rules were reinforced for non-religious reasons: agricultural ('by the eating of fish, much flesh is saved to the country'), economic (the fishing industry was encouraged) and political (a good fishing industry provided expertise and manpower for the navy and for voyages of exploration).

It was perfectly possible, of course, to obey the letter of the meatless law without actually being abstemious if one was so inclined. There was nothing *maigre* about some of the grand fish dishes produced for fast days. The author of *The Compleat Cook* (1658) gave a recipe for carp pie enriched with the blood of the carp and the flesh of a fat eel, and ended with the comment that 'this is meat for a Pope'. A French cookbook of 1702 provided the following recipe—surely the ultimate in fish pie exotica—but nevertheless quite correct for a day of abstinence:

A Pan-pie made of Carps-roes and Tongues

The Tongues and Roes of the Carps must be laid in order upon a piece of fine Paste, in the bottom of the Pan; season'd with Pepper, Salt, Nutmeg, fine Herbs, Chibbols, Morilles, common Mushrooms, Truffles and sweet Butter. Then, all being cover'd with a Lid of the same Paste, let the Pie be bak'd with a gently Fire, and serv'd up with Lemmon-juice.

During the 40 days of Lent in the Christian calendar, fish was also forbidden, and at certain periods of history, so were dairy products and eggs. Pie-cooks of the time became very creative with substitutions such as almond milk and rice, nuts and fruits—and the corpus of pie recipes is the greater for their efforts.

Sweet Pies

If you are one of those sweet-toothed souls to whom 'pie' means a dessert, you have probably been frustrated up to now as meat pies have dominated our story. Your time is now come.

Fruit pies started to come into their own during the sixteenth century as sugar became cheaper and more delicate forms of pastry were available. It is not that fruit was absent from pies before this time—far from it—but it was rarely a primary ingredient. The first predominantly fruit pies were still called 'bake-metes' ('meat' in its old sense of any solid food), but these were not fruit pies as we now know them, and not just because of the thick coffin-crust. One medieval recipe for a 'bake-mete' of pears instructs that 'gobbets of marrow' (bone marrow, not vegetable) be placed between the pieces of fruit and in another the apple is flavoured with saffron:

For To Make Tartys In Applis

Tak gode Applys and gode Spycis and Figys and reysons and Perys and wan they are wel ybrayed colourd wyth Safroun wel and do yt in a cofyn and do yt forth to bake wel.

There is one other historic fruit pie that deserves special mention: a pie made from an exotic, imported and therefore expensive fruit candied with expensive sugar. Orengeado is candied orange peel, and it was enormously popular from Elizabethan times until well into the eighteenth century. A pie made from orengeado, perhaps layered with apples, was a very expensive delicacy. A pie fit for a queen, it seems, as two of the master pastry-cooks of Elizabeth I were proud to make New Year gifts of an orengeado pie to her in1600.

We have been careless with our pie repertoire. The demise of apple-pear pie with figs and saffron and orengeado pies are tragic losses. What did we replace them with? Dessert pies have moved well beyond fruit and custard, and the line is blurred between pies and cakes with some pies resembling cakes with a crust (pecan pie springs to mind). Some sweet pies are even made with vegetables.

Vegetable Pies

Vegetable pies are not absent from historic cookbooks, but the importance and role of vegetables in the diet has changed significantly over the centuries. Aside from the members of some religious orders who abstained from it, our medieval ancestors would have found it incomprehensible that an individual would freely choose not to eat meat. High-protein food was too hard to come by, especially in the European winter, and there was never too much.

There are some fine vegetable pies in old cookbooks, although as with fruit pies they are not necessarily strictly 'vegetarian' (a relatively modern word), often containing marrow from bones where we might use butter, such as in this recipe from about 1720.

An Artichoke Pye

Take ye bottomes of 6 or 8 artichokes being boyld & sliced season ym wth sweet spice mix ym wth ye marrow of 3 bones wth citron & lemon piele oringoe roots damsons gooseberries & grap[e]s citron lemon butter & close ye pye: A Skarrot or a Potatoe pye is made ye same way.

Vegetable pies were not always 'savoury' dishes. The traditional Thanksgiving pumpkin pie has an old lineage. There are seventeenth-century recipes for 'pompion pie' (pompion is an old name for pumpkin) and other intrinsically sweet vegetables such as sweet potato and skirret (a member of the carrot family) have long been used as pie fillings. The sweet–savoury combination applied to 'vegetable' pies too. William Ellis's farming and household manual from 1750 has a recipe for an 'onion pie' that is half onion and half apple, with no indication as to which ingredient was the substitute for a dearth in the other, and no clue as to when it was eaten.

Those who do not choose vegetarianism (or veganism, an even newer word) may have it thrust upon them. Peasants in most societies, the faithful during Lent and the general population during wartime do not choose. Cooks have two basic options when faced with absent or forbidden ingredients: embrace the alternative wholeheartedly and with creativity, or aim to make the substitute look and taste as close to the real thing as possible (and tell, or not.) During the Second World War in Britain, economies had to be made with wheat, and meat was rationed, so whither that staple of the working man's dinner, pie? Frederick Marquis, first Earl of Woolton, was appointed Minister of Food in Britain in 1940, and charged with the job of organizing the rationing programme. He managed a seemingly impossible feat by doing the job brilliantly and at the same time becoming enormously popular with the British public. A recipe for a vegetable pie to serve on meatless days was named after him, and described as being 'on the outside exactly like a steak and kidney pie, and on the inside just like a steak and kidney pie—without the steak and kidney'. The celebrity testimonial was a propaganda triumph rather than a culinary one, but do judge for yourself.

Woolton Pie

Take 1lb each diced of potatoes, cauliflower, swedes, and carrots, three or four spring onions—if possible, one teaspoonful of vegetable extract, and one table-spoonful of oatmeal. Cook all together for 10 minutes with just enough water to cover. Stir occasionally to prevent the mixture from sticking. Allow to cool; put into a piedish, sprinkle with chopped parsley, and cover with a crust of potato or wheatmeal pastry. Bake in a moderate oven until the pastry is nicely browned and serve hot with a brown gravy.

Vegetable pies do not need to be apologetic alternatives to meat, however, as the slightly mysterious Frenchman Baron Brisse in his cookbook of 1868 showed with a recipe for a vegetable pie in the form of an open tart divided into sections by strips of paste, each section filled with a different-coloured vegetable. As he pointed out, this 'not only pleases the palate, but the eye, and is a great addition to our fast day dinners'.

Frugal Pies

The domestic art of economy has been prized by cookbook writers for centuries. It was summed up nicely by the American author Lydia Maria Child in her book *The Frugal Housewife: Dedicated to Those Who Are Not Ashamed of Economy* (1830):

> The true economy of housekeeping is simply the art
> of gathering up all the fragments, so that nothing be lost. . .
> Nothing should be thrown away so long as it is possible to
> make any use of it, however trifling that use may be.

By their very nature pies are an eminently suitable device for practising this art, but it can perhaps be taken too far, as Charles Dickens graphically described in *Our Mutual*

Friend:

> When dinner was done, and when what remained
> of the platters and what remained of the congealed gravy
> had been put back into what remained of the pie, which
> served as an economical investment for all miscellaneous
> savings, Riderhood filled the mug with beer and took a
> long drink.

Dickens's American contemporary, Nathaniel Hawthorne was not too sure about pies either, at least the English hotel version, where 'sometimes, perhaps, a meat-pie, which, if you eat it, weighs upon your conscience, with the idea that you have eaten the scraps of other people's dinners'.

Pies such as these—repositories of a week's leftovers—were once so commonplace as to earn their own names. I advise you to have no illusions as to the content of Scrap pies, Saturday pies or Old Maid pies.

Frugality (short of parsimony, that is) may well be a domestic virtue, but there have always been many for whom it is a necessity. Surely the most frugal pie of all, for the poorest folk of all, was the Burr Pye. An early eighteenth-century manual 'for the improvement of husbandry and trade' described the butchering of an animal and the sending of the hides to market 'where poor women . . . also cut off some bits of flesh that lie by the horns, called *burrs*, with these are made pyes.'

Sometimes frugality must be practised on a national

scale. During the Second World War there was a drive in the United States to cut wheat consumption by 40 per cent and fat by 20 per cent, and one of the requests made of American housewives was that they made 'coverless' pies. The severe wheat shortages that occurred in England in the eighteenth century had a couple of interesting long-term results. The pie, as we have seen, was important for its 'presence' on the dining table. To compensate for their lack when wheat was scarce the great pottery manufacturers designed dishes with a golden glaze and scalloped edges to mimic the structure of a pie—so-called piecrust ware. These 'crock pies' were early versions of pot pies. Another positive outcome was that the crustless fillings developed independent lives and went on to be served as terrines, puptons and pâté.

Cheating Pies and Sinister Pies

There is a mystery inherent in a pie by virtue of its contents being hidden beneath its crust. The mystery is both its promise and its curse. We want a pleasant surprise, not a nasty shock. The big question, when the pie is broached, is whether we are opening Pandora's box or a treasure chest. Meat pies are the real worry here. The meat pie is the point at which many of the fine lines between frugality, harmless deception and sinister intent can meet. Happily, the law is there to protect us.

Australian law says that a meat pie must contain a

minimum of 25 per cent meat, which hardly sounds sufficient. The real anxiety starts when you know that this can include snouts, ears, tendons, blood and blood vessels and a whole lot of other animal parts (but not foetuses or offal, which is possibly reassuring). In fact, as the method of testing for meat actually tests for protein, it would be perfectly possible for a pie made of congealed thickened blood to pass the meat test.

There is a different set of motives at work at the domestic level when a host serves up one food as another, perhaps with the intent to deceive. Presumably this is intended to impress the guest, in which case the deception had better be good and remain undiscovered. Samuel Pepys dined at the home of his cousin on 6 January 1660 and was clearly unimpressed as 'the venison pasty was palpable beef, which was not handsome'. On another occasion he ate a venison pasty 'which proved a pasty of salted pork'. Cookbooks of the seventeenth and eighteenth centuries frequently contained recipes for cooking beef to mimic venison, so one has to assume that this was common practice at the time.

Fruit pies can deceive too. One does hear occasional rumours in Australia of 'apple' pies being made entirely from choko (chayote), for example, but at least the substitute is also from the plant kingdom. For sheer marketing nerve it is hard to beat the Ritz cracker manufacturers in the 1930s, who for many years proudly displayed a recipe for Mock Apple Pie on the package. The pie filling was made entirely from crackers, and difficult though it may be to believe, it was popular to the extent of becoming almost a cult item. It is even more difficult to believe the many fans who swore it was indistinguishable from the real thing.

The idea of fake fruit pies pale into insignificance in comparison to that of meat pies made from diseased or fake meat, and there is worse in the world than 'meat' pies filled with snouts and tails. And I don't mean the 'covered uncertainties' with the 'feline character' sold by street vendors in Victorian England and described by Albert Smith in *Sketches of London Life and Character* (1849). There is something much worse called called 'mechanically retrieved meat'—that is, 'meat' scraped off carcasses and used in burgers and pies. A sort of reddish paste suspected of spreading Mad Cow Disease—and you don't get much more sinister than that. Or do you?

A primal fear—one which has fed many a literary tale and many an urban myth—is that of being fed human flesh in a pie. Shakespeare used this gruesome idea as the ultimate weapon of revenge in his extraordinarily bloodthirsty play, *Titus Andronicus*. Titus is a Roman general who kills the remaining two sons of his enemy Tamora, Queen of the Goths, and has their flesh made into 'two mighty pies' which unknowingly, she eats. The tale of Sweeney Todd, the London barber who slits the throats of his customers and has them made into pies by his paramour to be sold to her own customers, has horrified and thrilled readers and audiences for a century or so, and may be loosely based on a factual murder that took place in 1785 in the same city.

History is littered with stories of cannibalism during times of great hardship and famine, and as much as we are revolted by the idea, we are also fascinated. What does human flesh taste like? Pork, the myths say. Perhaps this is why pork was used in the 'mermaid pie' that was so popular in the seventeenth and early eighteenth centuries. Mermaids were a source of great fascination at that time, with great scientific interest in supposed sightings. There was a vigorous debate among some clergy as to whether or not eating mermaids (who were half human after all) amounted to cannibalism. A recipe for mermaid pie appears, essentially unchanged, in cookbooks for over a century. Here is one version, from William Salmon's *The Family-Dictionary, or, Houshold Companion* (1695):

Mermaid-Pye

Take a Pig, scald it, and bone it; and having dried it well with a Cloath, season it with beaten Nutmeg, Pepper, and chop'd Sage; then take two Neats-Tongues; when dried and cold after boiling, and slice them in lengths, and as thick as a Half-Crown, and lay a quarter of your Pig in a square or round Pye, and the slices of the Tongue on it; then another quarter, and more Tongue: and thus do four times double, and lay over all these some slices of Bacon, scatter a few Cloves, put in some pieces of Butter and Bay-leaves, then bake it; and when it is so, fill it up with pieces of sweet Butter, and make your Past white of the Butter and Flower. This Pig, or Mermaid-Pye, so called, is to be eaten cold.

On an entirely different note, there is a type of pie strongly associated with Scotland which has aesthetic and health dangers that justify its inclusion here amongst the sinister pies. It is the Fried Pie—which is just what it says, a baked pie cooked a second time by frying. Scotland is not called the Land of the Brave for nothing.

Chapter 5   Special Occasion Pies

Methinks a Feast is not well set forth if there bee no Pies or bak'd meates.

John Taylor (1578–1653)

Practical, versatile, universally esteemed and provided with its own edible, easily decorated gift box of pastry—small wonder that the pie still plays a feature role at many of our favourite celebrations, so much so that it is often symbolic of the very event itself.

Christmas Pie

In the beginning, there was frumenty—a plain wheat porridge that was the staple food of peasants and a side dish to venison for the rich. It was enriched for special occasions (such as Christmas) with sugar and spice and all other things nice, such as eggs, dried fruit ('plums'), wine and finely chopped meat. This Christmas porridge (or pottage) eventually evolved into Christmas (mincemeat) pie when it was cooked in a coffin, Christmas pudding when it was cooked in a cloth and Christmas cake when it was cooked in a shaped metal tin.

The mince(meat) pie may have lost its meat, and its other ingredients may now be freely available all year round, but it has not lost its association with Christmas. Seventeenth-century Puritans tried hard to ban it (calling it 'idolatrie in crust') but they did not succeed: the Christmas mince pie lives. Not so another famous Christmas pie—a grand pie, a pie in the coffin style, a pie solid with the finest meat. The most famous of all came from Yorkshire, and the earliest recipe is from Hannah Glasse's *Art of Cookery*:

To make a Yorkshire Christmas Pye

First make a good Standing Crust, let the Wall and Bottom be very thick; bone a Turky, a Goose, a Fowl, a Partridge, and a Pigeon, season them all very well, take Half an Ounce of Mace, and Half an Ounce of Nutmegs, a Quarter of an Ounce of Cloves, and Half an Ounce of Black Pepper, all beat fine together, two large Spoonfuls of Salt, and then mix them together. Open the Fowls all down the Back, and bone them, first the Pigeon, then the Partridge, cover them; then the Fowl, then the Goose, and then the Turky, which must be large: season them all well first, and lay them in the Crust so as it will look only like a whole Turky; then have a Hare ready cased, and wiped with a clean Cloth. Cut it to Pieces, that is, joint, season it, and lay it as close as you can on one side, and on the other Side, Woodcock, more Game, and what Sort of wild

Fowl you can get. Season them well, and lay them close; put at least four Pounds of Butter in the Pye, then lay on your Lid, which must be a very thick one, and let it be well baked. It must have a very hot Oven, and will take at least four Hours.

This Crust will take a Bushel of Flour; in this Chapter, you will see how to make it. These Pies are often sent to London in a Box as Presents; therefore the Walls must be well built.

Thanksgiving

America has developed a pie tradition unequivocally and unapologetically at the sweet end of the scale, and at no time is this better demonstrated than at Thanksgiving in November. It seems that the country goes pie-mad at this time, and the traditional pies reflect that this is harvest season. Regional differences are inevitable of course, and the food writer Clementine Paddleford claimed to have summarized them in the 1950s when she said, 'Tell me where your grandmother came from and I can tell you how many kinds of pie you serve for Thanksgiving.' If she was from the Midwest, Ms Paddleford said, there would be two types (mince and pumpkin), if from New England, three (mince, pumpkin, cranberry), Boston, four (mince, pumpkin, cranberry and apple). Ms Paddleford excluded the West for reasons she did not explain, and the South on the rather controversial basis that pie was not favoured there.

There is no mention of savoury pies anywhere in any discussion of Thanksgiving. The American preoccupation with sweet dessert pies is absolute—an interesting point which we will explore further in chapter Six.

Village Pies

Every English village seems to have a particular day when some local tradition is celebrated—and the celebratory food is often pie. A few stellar examples must suffice. On Easter Monday in the village of Hallaton in Leicestershire, for ancient and forgotten reasons, the local parson is obliged to provide his congregation with a hare pie after the morning service. The remains of the pie are then taken to a field called Hare Pie Bank on the outskirts of the village, to be tossed to the crowd before the unruly traditional game of bottle-kicking begins.

In the fishing village of Mousehole in Cornwall it is traditional to eat 'stargazy pie' on the evening of 23 December. It is an intriguing pie, made with pilchards placed so that their heads poke through the crust at the centre of the pie, gazing at the stars, as it were. It is made in honour of a local mythical hero, Tom Bawcock ('bawcock' is an old word meaning 'a fine fellow'), whom legend says set out on a bad night during a bad season, returning with sufficient fish to save the locals from starvation.

The folk of Lapford in Devon used to tuck into 'pestle pies' on the feast day of the Translation of St Thomas Becket (7 July). A 'pestle' is the leg (thigh) of an animal, and a pestle pie is a large 'standing' pie which 'contains a whole gammon, and sometimes a neat's tongue also, together with a couple of fowls, and if a turkey not the worse'. The local church dedicated to the saint was built in the twelfth century by a local nobleman as penance for his part in the Arch-bishop's murder in 1170, but how this is connected to gammon pies is, to say the least, obscure.

Obligatory Pies

It was not uncommon in times gone by for items of food to be given as symbolic gifts or token payments, and pies were ideal as they could be impressive, extravagant and would keep and transport well.

The city of Gloucester, by ancient custom, presented a lamprey pie to the sovereign at Christmas time, as a token of loyalty. Lampreys are scaleless freshwater sucker-fish resembling eels, desirable in the past for their oily, gamey flesh. The tradition of gifting lamprey pies to the royal family continued until the end of Queen Victoria's reign, but was revived for the coronation of Queen Elizabeth II in 1953 when a 42-pound pie was cooked by the RAF catering corps.

The number 24 pops up regularly in pie history, and folklorists continue to debate the meaning of the nursery rhyme 'four-and-twenty blackbirds baked in a pie'. A 'double dozen' is a significant number in many cultures, not least as the number of herring pies present on symbolic gifting occasions. The reasons behind this are difficult to explain.

The choice of herring pies themselves is more obvious: they were desirable because they kept well, were rich (oily) and were suitable for fast days. The town of Yarmouth, famous for its herrings, was required by ancient charter to send 24 herring pies to the king each year, and the Sheriffs of Norwich had the same obligation at Lent to their local Lord of Caister.

A final example is of an individual obligation. A shepherd family called Edwards lived for centuries in a small cottage in the parish of the church of St Constantine in Harlyn, Cornwall. The annual rent was a single limpet pie (with raisins and herbs), due on 9 March, the feast day of the saint. It was a cheap rent indeed as limpets were food for the poor—freely available from the tide-line.

Bride Pie

Before there was wedding cake, there was bride pie. Bride pie did not give rise to the modern wedding cake, which materialized from the same path that led to Christmas cake. Bride pie was an entirely different entity. Initially there was no specific recipe and there were no mandatory ingredients; it was simply a pie containing the best delicacies that could be procured at the time. Such a pie was often called a 'batalia pie', the name coming from *beatilles*, meaning beautiful little things—things such as cocks' combs, lambs' stones (testicles) and goose giblets. The origin of the name was clearly lost upon cooks, who thought that it referred to battlements, and consequently often constructed batalia pies in the form of castles, complete with turrets. With or without turrets, it was a pie entirely suitable for a celebration, and some early recipes for batalia pie specified that it was suitable for a bride's pie.

The most amazing bride's pie recipe comes from the cookbook of Robert May, *The Accomplish't Cook*. The seventeenth century was the pinnacle of English pie-making, and this pie was the absolute zenith:

To make an extraordinary Pie, or a Bride Pie, of severall Compounds, being several distinct Pies on one bottom.

Provide cock-stones and combs, or lamb-stones and sweet-breads of veal, a little set in hot water and cut to pieces; also two or three oxe pallets blanched and slic't, a pint of oysters, sliced dates, a handful of pine kernels, a little quantity of broom-buds pickled, some fine inter-larded bacon sliced, nine or ten chestnuts roasted and blanched, season them with the salt, nutmeg, and some large mace, and close it up with some butter. For the caudle, beat up some butter, with three yolks of eggs, some white wine or claret wine, the juyce of a lemon or two, cut up the lid, and pour on the lear, shaking it well together, then lay on the meat, slic't lemon, and pickled barberries, and cover it again, let these Ingredients be put into the middle or scollops of the Pie.

Several other Pies belong to the first form, but you must be sure to make the three fashions proportionably answering one the other; you may set them on one bottom of paste, which will be more convenient; or if you set them several you may bake the middle one of flour, it being baked and cold, take out the flour in the bottom, and put in live birds, or a snake, which will seem strange to the beholders, which cut up the Pie at the table. This is onely for a Wedding to pass away time.

Now for the other Pies you may fill them with several Ingredients, as in one you may put oysters, being parboild and bearded, season them with large mace, pepper, some beaten ginger, and salt, season them lightly, and fill the Pie, then lay on marrow and some good butter, close it up and bake it. Then make a lear for it with white wine, the oyster liquor, three or four oysters bruised in pieces to make it stronger, but take out the pieces, and an onion, or rub the bottom of the dish with a clove of garlick; it being boild, put in a piece of butter, with a lemon, sweet hearbs will be good boild in it, bound up fast together; cut up the lid, or make a hole to let the lear in, & c.

Another you may make of Prawns and Cockles, being seasoned as the first, but no marrow: a few pickled mushrooms (if you have them) it being baked, beat up a piece of butter, a little vinegar, a slic'd nutmeg, and the juyce of two or three oranges thick, and pour it into the Pie.

A third you may make a Bird Pie; take young Birds, as larks, pulled and drawn, and a force meat to put in the bellies made of grated bread, sweet herbs minced very small, beef suet, or marrow minced, almonds beat with a little cream to keep them from oyling, a little parmisan (or none) or old cheese; season this meat with nutmeg, ginger, and salt; then mix them together with cream and eggs like a pudding, stuff the larks with it, then season the larks with nutmeg, pepper, and salt, and lay them in the Pie, put in some butter, and scatter between them pine-kernels, yolks of eggs, and sweet herbs, the eggs and herbs being minced very small; being baked make a lear with the juice of oranges and butter beat up thick, and shaken well together.

For another of the Pies, you may boil artichocks, and take onely the bottoms for the Pie, cut them into quarters or less, and season them with nutmeg. Thus with several Ingredients you may fill up the other Pies.

Weddings inevitably lead to families, and many family events in the past were deemed worthy of a special pie (or cake). We no longer make 'groaning pies'—the pies to feed attendants and visitors at 'groaning time' (an old but apt name for childbirth) or christening pies, but 'funeral pies' are still known amongst the Pennsylvania Dutch. They are otherwise known as raisin (or rosina) pies, on account of their main ingredient, chosen perhaps because they are always available, and the pies travel well.

Giant Pies

Culinary history is littered with stories of supersize pies, and many of these have been Christmas pies. If you thought Mrs Glasse's pie was big, how about the one prepared in 1826 by Mr Roberts, victualler in Sheffield? It was 'composed of rabbits, veal, and pork, in such quantities, as to weigh, before being carried to the oven, 15 stones 10 pounds'. An even bigger one was made in 1835 by Mrs Kirk, of the Old Ship Inn, Rotherham, 'which when taken to the oven, weighed upwards of seventeen stone; it consists of one rump of beef, two legs of veal, two legs of pork, three hares, three couple of rabbits, three geese, two brace of pheasants, four brace of partridges, two turkeys, two couple of fowls, with 7½ stone of best flour.'

The citizens of the tiny Yorkshire village of Denby Dale know a thing or two about giant pies: they have been making them for over 200 years. The first was made in 1788 to celebrate the recovery of King George III from one of his bouts of madness. The second was made in 1815 to celebrate the victory at Waterloo in 1815 and the third in 1846 to celebrate the repeal of the Corn Laws. Two pies were made in 1877 for Queen Victoria's jubilee. The first (made by professional bakers), which weighed one and a half tons, was embarrassingly unfit for consumption due to the meat being 'off ', and was quickly buried; the 'Resurrection Pie' was made (by local housewives) a month later. In 1928 a pie was made to raise hospital funds; this one was 16 feet long by 5 feet wide by 15 inches deep; the pastry required 1,120 pounds of flour and 200 pounds of lard, and there were 1,500 pounds of potatoes and the meat of six bullocks in the filling. The 1964 pie was also a fundraiser, to build a village hall. The Denby Dale Bicentenary Pie, made in 1988, measured 20 feet long, 7 feet wide and 18 inches deep. The most recent pie in 2000 was a triple celebration: the 12-tonne Millennium Pie also acknowledged the Queen Mother's 100th birthday and the 150th anniversary of the local railway line.

Entertainment Pies

It could be argued that there is an element of entertainment in every pie, as every pie is inherently a surprise by virtue of its crust. The medieval 'chastlete'—the pie in the style of a castle (page 48) would have been great fun—and great eating—but there are some pies in an entirely different league that are clearly meant to entertain, and just as clearly *not meant to be eaten*. These pies depend for their success on the anticipatory delight that is the right of every pie-eater. A sixteenth-century Italian recipe will serve to illustrate:

To make Pies that the Birds may be alive in them, and flie out when it is cut up

Make the coffin of a great pie or pastry, in the bottome thereof make a hole as big as your fist, or bigger if you will, let the sides of the coffin bee somewhat higher then Ordinary pies, which done put it full of flower [flour] and bake it, and being baked, open the hole in the bottome, and take out the flower. Then having a Pie of the bigness of the hole in the bottome of the coffin aforesaid, you shal put it into the coffin, withall put into the said coffin round about the aforesaid pie as many small live birds as the empty coffin will hold, besides the pie aforesaid. And this is to be done at such time as you send the Pie to the table, and set before the guests: where uncovering or cutting up the lid of the great Pie, all the birds will flie out, which is to delight and pleasure shew to the company. And because they shall not bee altogether mocked, you shall cut open the small Pie, and in this sort you may make many others, the like you may do with a tart.

This idea of live birds flying out of a pie to give delight and pleasure to the guests was popular for centuries. Robert May (the author of *The Accomplish't Cook*) went one better with instructions for an incredibly complex pie/pastry construction that also contained frogs which, when released, would 'make the ladies to skip and shreek'.

The concept was taken several leaps further at a time not at all concerned with political correctness. Jeffrey Hudson was born in 1619, and at the age of nine years he stood only

18 inches tall (although 'gracefully proportioned'.) He went on to live an incredibly exciting life, fighting duels, being captured by pirates and spending time as a political prisoner, but he first came to fame as the surprise ingredient in a pie in 1626. The lucky and very amused recipients were Charles I and his wife Henrietta Maria, who immediately 'adopted' (i.e., made a pet of) him.

Another interpretation of the same theme occurred in New York in 1895. A select group of gentlemen in that city were invited to a secret dinner and a mysterious treat was suggested. The treat turned out to be a pie containing a cloud of canaries—quickly followed by sixteen-year-old Susie Johnson, wearing nothing very substantial. What happened next has never been authenticated but, needless to say, the general public (who were not invited) were outraged at the whole concept of a 'Pie Girl Dinner'.

Chapter 6   Around the World with Pie

Do not dismiss the dish saying that it is just, simply food.

The blessed thing is an entire civilization in itself.

Abdulhak Sinasi

We humans are constantly on the move around the world and when we migrate we take our eating habits with us. We do so to use our agricultural and culinary knowledge, and because eating familiar food maintains our link with home and eases our homesickness. We may have to substitute ingredients and adapt our cooking methods, but even after several generations, our heritage is still evident in the food we serve at home.

When Europeans started to spread beyond their traditional homelands, they took their grain-based cuisine with them, and attempted to make familiar types of bread and pastry. By the seventeenth century, England was at its peak in both its maritime powers and its culinary skill, and as England's empire developed, its pies went with it, to be adapted according to local ingredients and conditions.

The two great waves of migration from pie-loving

Britain—to the Americas starting in the early seventeenth century, and to Australia in the late eighteenth century—produced two great pie-loving nations. What is interesting is that, in spite of this common origin, the pie portfolios of these two ex-colonies are very different. In America, the unqualified word 'pie' unequivocally means a sweet dessert item, whereas in Australia it just as certainly means a *meat* pie.

Nowhere is this divide better demonstrated than in the pie competitions in both countries. The American Pie Council's Annual National Pie Championship has twelve to fourteen categories, all sweet (always apple, fruit/berry, cream, citrus, custard, pumpkin, sweet potato, nut, chocolate cream and peanut butter, plus a variable few others.) By way of contrast, the Great Aussie Pie Competition (for commercial bakers) in 2006 had five categories only: red meat, poultry, game, fish, vegetarian—not a spoonful of sugar anywhere.

So, we have two countries divided by a common culinary heritage. How might this have come about?

America

The first mass migrants to North America—the Puritans and Pilgrims—arrived with the intention and equipment to be small farmers, and the area they chose to settle proved to be excellent for their purpose. In particular the apple so beloved in old England grew wonderfully well in New England, and within a few decades it was a staple. Dried apples became a standard provision for the long overland journeys to the West, and were used for one major purpose—apple pies.

It was a different story with the wheat crops in those early years. As the early settlers finally learned from the Native Americans, this was corn (maize) country, and not suitable for wheat. Wheat was not grown on a massive scale until the West was well and truly won, by which time Americans had learned to be sparing with it—and a small amount of wheat goes further if it is used to make pies rather than bread.

It is hardly surprising that to this day New England is considered to be the pie capital of America, whose inhabitants traditionally eat (sweet) pie for breakfast. Apple pies in particular became deeply embedded in the history of America—associated with the old country, the new country and the pioneering spirit, and indelibly identified with the sense of nationhood and patriotic sentiment.

Australia and New Zealand

The first Europeans to make their home in Australia did so, for the most part, unwillingly. They were convicts and marines and were largely devoid of agricultural or culinary skill, coming as they did predominantly from the urban poor of the large cities spawned by the Industrial Revolution. For the urban poor, meat was a prize, and if they had it at all it was likely to be in the form of a pie bought (or stolen) from a cookshop or street trader. These first migrants came to a country ideally suited for wheat-growing and grazing, and indeed the later wave of free settlers were lured to the colony with the promise of 'meat three times a day'.

The Lord-Mayor elect of Sydney in 1972 is quoted as saying that Australia was built on 'meat pies, sausages and galvanised iron', and there is a satisfying irony in the fact that, in at least one example, he was quite literally correct. When the new Parliament House in Canberra was opened in 1927, the general public was underwhelmed by the whole event, forcing caterers to bury a massive amount of prepared food (including 10,000 pies) in the local rubbish tip. Later the same year, an administration building was erected on that particular piece of ground. The building now houses the Department of the Environment and Water Resources—which somehow seems very apt, particularly when you know that, after its completion, some 620 tons of cement intended for use in its foundations was discovered unused. The stale meat pies must have been deemed sufficient.

It seems that the country's early love affair with the meat pie was shared across all ranks. The first recorded mention of a pie in an Australian newspaper was in the Melbourne *Argus* in 1850, in an article which noted that the town councillors preferred meat pies from the local pub to the food provided in the council chambers.

Unofficial regard is one thing, but one particular style of pie called the 'pie floater' is officially recognized as a

South Australian Heritage Icon by the National Trust of Australia. A fine symbol, many would say—a meat pie served in a puddle of mushy peas, garnished with tomato sauce. It is indeed symbolic, if Terry Pratchett has summed it up correctly in his 'vaguely Australian' story, *The Last Continent*: 'Who is this hero striding across the red desert? Champion sheep shearer, horse rider, road warrior, beer drinker, bush ranger and someone who'll even eat a Meat Pie Floater when he's *sober*?' The affectionate esteem (accompanied as it is with a total lack of illusion) in which pies are held in Australia is best reflected in some of the slang expressions for pie. Anyone for a fly cemetery, rat coffin or maggot bag?

Australians like to fancy that they are the greatest pie consumers in the world, but that accolade may well fall to New Zealanders. Exact figures are impossible to determine—large commercial bakeries are shy of releasing their information, and mom and pop bakeries never do— so the prize may never be formally awarded. As in Australia, meat not sweet pies rule in New Zealand, and they are the traditional fast food at sporting venues. Naturally, given the country's extreme suitability for sheep-rearing, the iconic pie in New Zealand is made from mutton.

Canada

Perhaps more than for any other modern culture, Canada's heritage and history is clearly revealed through its pies. The Cape Breton 'pork' pie, porkless though it may now be, is clearly descended from the medieval European tradition. Did it come via the Scots, who escaped there in huge numbers from the dramatic social and agricultural changes of the first half of the nineteenth century? Or from the French, who may have given the island its name? Brittany (inhabited by Bretons) is an area of France famous for its butter (compared with the rest of the country where oil dominates) and pork.

Canada's French heritage is reflected linguistically at least, in two popular Quebecois pies; the *tourtière*, a double-crust meat pie especially popular on Christmas Eve, and the *cipaille* or *cipate*. The *cipaille* is a multilayered pie of meat, potatoes and pastry (the name may derive from 'six pâtés'), which is strangely similar in composition and sound to a well-known English dish called 'sea pie'.

Finally, there is one pie in Canada which is unequivocally its own, born of the hard life of Nova Scotia. It is the 'seal flipper pie'. In the early days, the seal fishermen of Nova Scotia had, in the usual way of such folk, to sell the best parts of their harvest, keeping only the unwanted scraps to feed their own families. Thus was the seal flipper pie born. It is, they say, an acquired taste.

Other Places

The pie has an indisputable European heritage, but

the areas in which oil is the dominant fat have by necessity taken their cuisine in other directions. Italy has the pizza. It may be correct that the word 'pizza' is usually translated as 'pie', but it must be understood that pizza is not pie, for the simple reason that bread is not pastry. If we allow pizza to be included as pie, we might as well call a toasted sandwich a pie, which would be ridiculous.

One does not think of pies (particularly meat pies) in association with classical French cuisine. France sits on the fat/oil cusp, with both being important in cooking, but butter being essential in classic sauces. Certainly in medieval times, food was cooked inside a crust, and we have seen how pâté refers to pastry, so that pâté de foie gras was formerly a pie of fatted goose liver. That is not to say that savoury pies cannot be found in France—there is the *quiche Lorraine* and the *tourte de la vallée* of Alsace, for example—but by and large pastries in France are of the small sweet variety and fill the same role as cakes.

The Germans and Austrians have a fine baking tradition, but they seem to have expended their energy in refining their prodigious cake repertoire. There does not seem to be an important pie tradition in those countries, where the potato is king of the savoury starches.

Further north, there are pies that reflect the cold, wheat-unfriendly climate and the strongly fish-based diet. These are the Finnish pies such as the Karelian pastries made mostly of rye flour, carbohydrate-heavy with fillings of barley or buckwheat or potato or rice, and eaten with melted butter and boiled eggs. On the same grounds as given for the pizza, the other Finnish 'pie' called *kalakukko* must be excluded as it is a rye bread 'pasty' filled with a mixture of pork, fish and bacon. Finland's old master, Russia, has its *kurniks*, or chicken pies, traditional on feast days, and *piroshki/pierogi*, closer to dumplings and deep fried or boiled.

There is one other type of pastry that we have not touched on so far, but which surely should not be neglected. It is filo pastry (the name refers to the fine 'leaves') so well known in Middle Eastern cuisine. Wrapped around a nutty filling and drenched in honey syrup, it is used to make some highly addictive pastry sweets such as baklava. Filo is not just used for tooth-achingly sweet pastries. There is one filo pastry pie that deserves special mention. It is the Moroccan pie called *b'stilla*, a sweet pigeon pie which is a legacy of the early Arab influence on pastry-making, and of the medieval tradition of sweet with meat.

Pasties

Today we think of a pasty as an eat-in-the-hand, single-serving savoury pastry in the 'turnover style', but it was not always so. A 'pasty' often meant something quite different in the past. There wasn't always a clear distinction between pies and pasties, but there was a tendency for pies containing a single piece of meat (especially venison) to be called a pasty. They were often huge. Samuel Pepys in the mid-seventeenth century refers several times to sharing a single venison pasty with several friends over several days. Mrs Mary Tillinghast in her book of *Rare and Excellent Receipts* (1690) says: 'If you make your Pasty of Beef, a Surline [surloin] is best; if of Mutton, then a Shoulder or two Breasts is the best. A Venison, or a Beef Pasty, will take six hours baking.' This was by no means the largest pasty—some recipes called for a bushel (36 litres volume) of flour, and up to 24 hours cooking.

Today's pasty is the working man's version, a perfect meal in the hand, easily transportable to the mines or the fields. Although they are indelibly associated with Cornwall, pasties can be found in one form or another all over Britain. One particularly superb example is the 'Bedfordshire clanger', from the English county of that name, which serves as both a main course and dessert by virtue of one end having a savoury filling, the other sweet.

The traditional ingredients of the 'oggie', as it is called in the old Cornish language, are naturally disputed, but on some things most experts agree: the meat must be chopped, not minced, the vegetables (perhaps potato, onion and turnip) must be sliced and the ingredients are not pre-cooked before they are put in the pastry. The twisted or 'crimped' edge is traditional too, and forms the 'handle' by which it is held for eating—hence the crimp must be on the side, not the top edge. The wives of Cornish tin-miners used to mark the pastry with their husband's initials to ensure that their man got the correct lunch, and they supposedly made the pasty sturdy enough to survive if it was dropped down the mineshaft.

The miners of Cornwall took their pasty tradition to America when they migrated in droves in the nineteenth century. The tradition is alive and well in the Upper Peninsula region of Michigan and in eastern Pennsylvania, and is one of the rare examples of a savoury pie tradition in the USA.

Chapter 7   Imaginary Pies

But I, when I undress me,

Each night upon my knees,

Will ask the Lord to bless me,

With apple pie and cheese.

Eugene Field

It would be foolish to say that art and literature offer us almost as great a feast of pies as has ever come out of our kitchens but, nevertheless, the pie-pickings are rich among the stories and pictures that form our cultural heritage. Our exposure to the imagery of the pie begins very early—before we ever eat them—in the form of the nursery rhymes that we learn to chant as children. Understanding is not necessary to the enjoyment of these apparently non-sensical jingles, but some infants grow up to become folklorists (and urban mythologists), certain that the rhymes of their childhood must have meaning and determined that it will be discovered and exposed. They are hampered by the fact that the oral tradition of these rhymes predates the written version, possibly by centuries in some cases. Inevitably then, any explanations offered involve significant conjecture.

Some interpretations are fanciful in the extreme, but usually they fall into one of two categories: they are based on historic events or people (and represent gossip or propaganda in a pre-mass media age), or they are cautionary tales for the instruction of children. The latter category probably includes the 'Three Little Kittens' (who were not going to get pie unless their lost mittens were found) and 'Simple Simon', who meets the pieman and learns, in spite of his simplicity, that a pie cannot be bought on credit.

There are two possible contenders for the real George in 'Georgie Porgie, Pudding and Pie', who kissed the girls and made them cry: George Villiers, second Duke of Buckingham (1628–1687) and the Prince Regent, the future George IV (1762–1830), both of whom were known to be amoral and gluttonous. Tradition says that 'Little Jack Horner' was based on a real person too. His real name was probably Thomas, and he was steward to the Abbott of Glastonbury during the period of the dissolution of the monasteries by Henry VIII. The 'plum' that he pulled out of the Christmas pie was the deed to the valuable manorial lands of Mells in Somerset. Depending on the version of the story you believe, he either stole the deeds which were in his safekeeping, or was given them in reward for services rendered. Of course it is entirely open to conjecture whether the pie itself was metaphorical or real—for, as we have seen, sometimes pies were used to hide non-edible materials.

So what of the blackbirds baked in the pie in 'Sing a Song of Sixpence'? There are at least half a dozen interpretations of this rhyme, some highly unlikely and some ridiculous, but it is possible that the origins of the story go back as far as 1454. In that year the Duke of Burgundy, Philip the Good, held a feast in Lille to whip up support for another crusade, and one of the entertainments was a giant pie containing a group of musicians who 'sang' when the pie was opened. The story says that there were 'eight-and-twenty' players, but a small difference in number does not destroy the theory altogether. Alternatively, the rhyme may be about the day—24 birds representing the hours, and the opening of the pie and the singing of the birds referring to the dawn. The rhyme is certainly *not* a coded message for the recruiting of pirates, as some urban myths insist!

Inevitably, as we leave our childhood we move away from nursery rhymes to books and films, but the pie continues to appear as character, plot device, prop or metaphor in our grown-up literature. If it has not already been written, there is an entire PhD in the use and significance of the pie in Charles Dickens's works alone. Sometimes his pies are substantial icons of hospitality (the huge breakfast Yorkshire pie in *Nicholas Nickleby*), often they are 'delusive' (the pigeon pie in *David Copperfield* that is 'like a disappointing head, phrenologically speaking: full of lumps and bumps, with nothing particular underneath'), and sometimes they are vaguely sinister, like the crust full of recycled table scraps in

Our Mutual Friend that we came across in chapter Four. The most delightful pie discussion in Dickens's work takes place in *The Pickwick Papers*:

> 'Weal pie,' said Mr Weller, soliloquizing, as he arranged the eatables on the grass. 'Wery good thing is weal pie, when you know the lady as made it, and is quite sure it ain't kittens; and arter all though, where's the odds, when they're so like weal that the wery piemen themselves don't know the difference?'

Mr Weller then goes on to repeat the advice from his pieman friend:

> '"It's the seasonin' as does it. They're all made o' them noble animals," says he, a-pointin' to a wery nice little tabby kitten, "and I seasons 'em for beefsteak, weal or kidney, 'cording to the demand. And more than that," says he, "I can make a weal a beef-steak, or a beef-steak a kidney, or any one on 'em a mutton, at a minute's notice, just as the market changes, and appetites wary!"'

A small meat pie provides more than basic nourishment in Francis Hodgson Burnett's *A Little Princess*. In the story there is an unsuitable friendship between Miss Sara and the ill-nourished, overworked little scullery maid, Becky. The kindly Miss Sara seeks out small edible treats for Becky which she hands over during their stolen moments together. On one occasion she brings some small meat pies, an amazing treat for the little maid, who enthuses, "'Them will be nice an' fillin'. It's fillin'ness that's best. Sponge cake's a 'evenly thing, but it melts away like—if you understand, Miss. These'll just stay in yer stummick.'" Becky's health visibly improves with the extra food, but she is sustained in more ways than one. She knows that 'the mere seeing of Miss Sara would have been enough without meat pies'.

The poet Louis Untermeyer wondered, 'Why has our poetry eschewed, The rapture and response of food?' Poetry has not completely eschewed the topic, but it does seem to have reserved the truly lyrical poems for luscious fruit, or very occasionally, as in the hands and heart of someone such as Pablo Neruda, for vegetables such as the tomato and onion. There are poems written about pies, but if they are not comical, the real subject is a larger theme. Robert Southey's 'Gooseberry Pie' is an ode to creation in general and to 'Jane' in particular; in John Greenleaf Whittier's poem 'Pumpkin Pie', the pie is symbolic of Thanksgiving; and the well-known quotation from the children's poet Eugene Field that forms the epigraph to this chapter comes from a poem that is a call to patriotism.

There are other opportunities for the pie to star once the medium becomes visual. They may not have featured frequently in movies, but when they do pies often steal the scene. Who could forget the role of the warm, homemade apple pie in *American Pie*? It must be the only film ever made in which the pie has a role as a sex-object.

An early (1901) silent movie called *Hot Mutton Pies* played on ethnic anxieties (and revenge) with its story of two men buying and eating a pie from a Chinese vendor who, to their horror, laughs and turns his sign around to reveal the words 'Alle Samee Cat Pies'. The recurring theme of anxiety on the part of pie-eaters about the contents of pie is exploited even more graphically in *Theatre Of Blood*, a 1973 horror movie starring Vincent Price. Price plays the part of a Shakespearian actor panned by the critics who gets his revenge by murdering each of them in a very individually appropriate way. The critic Meredith Merridew (played by Robert Morley), in a scene no doubt inspired by Shakespeare's *Titus Andronicus*, chokes to death when forced to eat a pie containing his 'children'—his beloved poodles. The even greater (almost primal) fear/ fascination of cannibalism is taken further in the 1992 film *Auntie Lee's Meat Pies*. The theme is not new; it is straight from the eighteenth-century story of Sweeney Todd. Auntie Lee's famous pies are made with the assistance of her four beautiful nieces who find her the filling material—handsome young men.

A discussion of the pie in movies would hardly be complete without mention of the classic comic device of custard-pie throwing, now legitimized and made semi-serious as the subversive political act of 'entarting'. 'Entarting' is delivering (by 'lovingly pushing', not throwing) a cream pie into the face of a deserving celebrity, preferably in full view of the world's media, in order to make a point. It happened to Bill Gates in 1984 at the hands of the famous

Belgian anarchist and *entarteur* Noel Godin, who considers the technique a form of communication—'a sort of visual Esperanto'. The actual message communicated does not always appear to be clear to the target or onlookers, but it is presumably so to the perpetrators, who claim to have been inspired by pie-throwing comedy movies. It is to be assumed then, that required viewing for a trainee *entartiste* would be the most spectacular pie fight in movie history—the four-minute sequence from the Laurel and Hardy film *The Battle of the Century*. More than 3,000 cream pies (the entire day's production of the Los Angeles Pie Company) were ordered and not one was wasted; every one found its target, and great fun was had by all.

Finally, let us not forget cartoons. Many cartoon characters have a favourite food (Garfield has lasagne, Dagwood has huge sandwiches, Jiggs has corned beef and cabbage) but for some (Popeye and spinach for example), a particular food is essential to the storyline. Desperate Dan, the cowboy superhero from the British comic *The Dandy*, has been 'the strongest man in the world' since 1937. His great strength is attributable to his love of cow pie, which is exactly what it sounds like—a pie containing an entire cow, horns, tail and all. Sadly, after seven decades, the source of his great strength is now off the menu lest it be thought that he is promoting mad cow disease.

On the animated graphics side, the pie-lover par excellence is the chief character in the Flash cartoon series

The Everyday Happenings of Weebl. The cartoon, which has almost a cult following, is minimalist in every way: Weebl and his friend Bob are simple egg shapes who roll around on a magenta background speaking in short phrases to the accompaniment of music, and their minimalist everyday routine usually revolves around the search for pie.

Epilogue: The Future of Pie

You can say this for ready-mixes—the next generation isn't going to have any trouble making pies exactly like mother used to make.

Earl Wilson

I asked at the beginning of this book, 'Is the pie dead?' We have better ways now to fulfil the original functions of the pie—we have baking dishes, lunch boxes, and refrigerators. We no longer *need* pies. But do we still *want* them?

The commercial pie seems to be alive, after a fashion. It has hung onto its old street-food role in the face of competition from the likes of the hot dog and the doner kebab, and in spite of there still being as much anxiety about its contents as there was in Chaucer's time. Some pies, such as Christmas mince pies and the Thanksgiving pumpkin pie have survived because they moved beyond mere tasty calories to become powerful symbols. The ordinary, everyday homemade pie has not been so successful or so lucky. Does it have a future in our rushed, pastry-phobic world? Pies are a labour (or a labour of love) to make, and we are all time-in-the-kitchen poor. On the other hand the conveniences of modern life mean we can buy frozen pastry and we don't even need to turn on the oven—we can cook our pies in bench-top pie-makers. Are we too lazy even for this?

If the pie dynasty is as I now suspect and fear, really dying, does it matter? Does anyone care? Should we simply look back with nostalgia and record its passing? Or should we try to save it? As is usual, such a question raises even more questions. I cannot save the seal and at the same time save seal-flipper pie. Would seal-flipper pie be worth saving even if it did not require the brutal deaths of baby seals? Is it right to try to save the glorious Yorkshire Christmas pie and ignore blood pie and horsemeat pie? I am pretty sure that I personally do not want to save Spam pie, but someone out else there probably does.

Surely we should try to save something that, when done well, is not only a supreme example of the art of cooking, but a dish that encapsulates humankind's entire culinary history?

Recipes

Pastry is one area of cookery in which accurate measurement is important. In earlier times—as Gervase Markham noted in 1615—it was assumed that the cook would know 'what paste is fit for every meate', and so a pastry recipe was not given for every type of pie.

The modern cook can get away with one recipe—for a basic shortcrust—for most pies. A few enthusiasts may tackle their own puff pastry but a good frozen brand is perfectly acceptable. Almost no one makes their own filo pastry, for obvious reasons, but almost everyone should try a hot-water crust at least once: it causes nowhere near as much anxiety as other forms of pastry.

Basic Shortcrust Pastry

The basic formula is half the amount (by weight) of fat to flour, with just enough water to bind, taking care to keep everything cool and handle as little as possible.

The following amount is sufficient for a double-crust

8-inch (20-cm) pie:

Ingredients

8 oz / 220g plain flour, sifted

a pinch of salt

4 oz / 110g fat (butter or a mixture of butter and lard)

2–4 tablespoons of cold water

Rub the fat into the flour and salt (or briefly pulse in a blender) until it resembles breadcrumbs. Add the water and mix quickly and lightly with the blade of a knife. When the dough starts to come away from the sides of the bowl, pat it gently together with your fingers. Wrap in plastic wrap and refrigerate for at least half an hour. Refrigerate again, if you can, for another half an hour after rolling and before cooking.

Hot-water Crust

This is suitable for the more robust type of meat pie or pasty. This amount is sufficient for four good-sized pasties.

Ingredients

10 oz / 400g plain flour and a good pinch of salt, sifted

4½ oz / 125 g lard or dripping

(can substitute half butter for flavour)

5 fl. oz / 150 ml milk or milk and water

Place the milk and fat in a saucepan and bring to the boil. Add to the flour and mix just until it is smooth. Cut into four and roll each out into a round (about ¼-inch or ½ cm thick) while it is still warm. Fill as desired.

Cornish Pasty

Ingredients

shortcrust or other pastry made with 1 lb / 450 g flour

¾ lb / 340 g beef cut into small dice (not minced)

1 onion, finely chopped

1 piece of swede or turnip (raw) cut into small dice

2–3 raw potatoes, cut into small dice

beaten egg for glazing (optional)

Roll out the pastry to about ¼ inch thick and cut into rounds (whatever size you wish). Mix the remaining ingredients gently and pile onto one half of each round (if you want the seal along the edge) or in the centre (if you want to make the seal along the top). Damp the edges and crimp together. Brush with the beaten egg if you wish to have a glazed finish. Bake in a moderate oven for 35–45 minutes.

Stargazy Pie

Ingredients

6 to 8 pilchards

1 medium-sized onion

3 rashers of bacon

1 lemon

2 free-range eggs

18 oz / 500 g shortcrust or flaky pastry

salt and pepper to season (sea salt for greater authenticity)

parsley and tarragon for flavouring

Cornish Fish Recipe Method

Gut, clean and bone the fish, leaving on the heads and tails (you may find the flesh is so fresh you can pull the backbone free without a knife). Finely chop the onion. Chop the bacon into squares. Cut the lemon in half; set two slices to one side for decoration. Squeeze and save the juice. Finely grind the rind. Boil the eggs until soft, then cut into small dice.

Cut the pastry mixture into two halves. Roll them out and place one half in an 8-inch shallow pie dish. Cut off the overlapped edges. Coat the edge with either milk or water to ensure the pastry lid will stick.

Then either: carefully place your pilchards into the bottom of the dish arranging them, like the spokes of a wheel, around the edge of your dish. Place the mixed chopped onion, eggs and bacon in the gaps between the fish. Some recipes suggest stuffing them with half the finely chopped mixture, but given the small gut of the pilchards, is it worth trying to do so? Add the lemon juice and cover with your pastry lid, pressing down around the fish to seal the pie, trim the edges of overlapping pasty and crimp the edges in true Cornish style.

Or (and we find this more authentic): place all your chopped ingredients, including seasoning, into the dish. Cover with the pastry lid, trim the edges of overlapping pasty, crimp as above, then carefully cut slits into the pasty, hold open with the blade of a knife and gently push the whole fish into the slots, leaving just the heads or tails showing. Add the lemon juice and then seal the slits. Coat the now completed pie with beaten egg.

Cooking your Pie

Place in the middle of a pre-heated oven (200℃) for around 30 minutes, until golden-brown. (For larger pies more time might be needed.) Serve piping hot with a sprig of parsley garnish and Cornish new potatoes.

Historical Recipes

It can be a frustrating experience for a modern cook to try to reproduce historic recipes. Old cookbooks are difficult to follow because they rarely contain exact measurements or temperatures, and the instructions are very vague. The older the cookbook, the more this applies—and often the old language is also incomprehensible. Cookbooks in medieval times were not primarily written as instruction manuals—cooks learned on the job, were not expected to need books and, in any case, were not necessarily literate. They were written as *aides-memoires* for others in the household who were responsible for provisioning the kitchen, or to be included in the master's library as an indication of his wealth. It was not until the mid-nineteenth century that cookbook instructions started to become more detailed and accurate: for most historic recipes we must make educated guesses and use our own judgement. I invite you to do this with the following recipes:

Tart de Bry

—from the Master-Cooks of King Richard II,

The Forme of Cury, c. 1390

Take a Crust ynche depe in a trape. take zolkes of Ayren rawe & chese ruayn. & medle it & þe zolkes togyder. and do þerto powdour gyngur. sugur. safroun. and salt. do it in a trape, bake it and serue it forth.

NOTE: 'Bry' appears to refer to the region where this cheese tart originated: the specific cheese mentioned is 'ruayn', which was a soft autumn cheese. Because the amount of sugar is not specified in the recipe we don't know how sweet this tart was intended to be—but as sugar was very expensive at the time, it seems reasonable to assume that this was closer to what we would now call a quiche, rather than a cheesecake.

For a tarte Of apples and orenge pilles

—from Anon., *The Good Huswifes Handmaid for Cookerie in Her Kitchen*, 1597

Take your orenges and lay them in water a day and a night, then seeth them in faire water and honey and let seeth till they be soft; then let them soak in the sirrop a day and a night: then take forth and cut them small and then make your tart and season your apples with suger, synamon and ginger and put in a piece of butter and lay a course of apples and between the same course of apples a course of orenges, and so, course by course, and season your orenges as you seasoned your apples with somewhat more sugar; then lay on the lid and put it in the oven and when it is almost baked, take Rosewater and sugar and boyle them together till it be somewhat thick, then take out the Tart and take a feather and spread the rose-water and sugar on the lid and let it not burn.

NOTE: This would have been a very extravagant dish. Oranges preserved in syrup ('orengeado') as in the first step of this recipe were a very expensive delicacy. A pie such as this—an 'orengeado pie'—was grand enough that two of her pastry-cooks each gave one to Queen Elizabeth I as a New Year gift in 1600.

An Herb Pie for Lent

—from Elizabeth Raffald, *The Experienced English Housekeeper*, 1769

Take lettuce, leeks, spinach, beets, and parsley, of each a handful. Give them a boil, then chop them small, and have ready boiled in a cloth one quart of groats with two or three onions in them. Put them in a frying pan with the herbs and a good deal of salt, a pound of butter and a few apples cut thin. Stew them a few minutes over the fire, fill your dish or raised crust with it, one hour will bake it. Then serve it up.

Pompkin [Pie]

—from Amelia Simmons, *American Cookery*, 1796

No. 1 One quart stewed and strained [pumpkin], 3 pints milk, six beaten eggs, sugar, mace, nutmeg and ginger, laid into paste No. 7, or 3, cross and chequer it, and bake in dishes three quarters of an hour.

No. 2 One quart of milk, 1 pint pompkin, 4 eggs, alspice and ginger in a crust, bake one hour.

No. 7 paste ['A Paste for Sweet Meats']

Rub one third of one pound of butter, and one pound of lard into two pound of flour, wet with four whites well beaten; water as much as necessary: to make a paste; roll in the residue of shortening in ten or twelve rollings—bake quick.

NOTE: This was the first cookbook to be printed in America. The recipe appears in her 'pudding' chapter. She also has a recipe for a 'potatoe' pie—using sweet potato.

Curry Pie of Fish or Meat

—from A Lady, *Domestic economy, and cookery, for rich and poor*, 1827

The curry ought always to be prepared for pies and cooled, or much better if dressed the day before, or any left curry will answer better than one newly made; put it into a nice puff paste covered with a thin cover, set round closely with long leaves, with the points upwards, and a deep border round, with leaves falling down from the top. A pastry formed in this way is very handsome.

Plain boiled rice and curry sauce, or curried rice, must be served with it.

A Good Apple Tart

—from Eliza Acton, *Modern Cookery in All its Branches*, 1845

A pound and a quarter of apples weighed after they are pared and cored, will be sufficient for a small tart, and four ounces more for one of moderate size. Lay a border of English puff-paste, or of cream-crust round the dish, just dip the apples into water, arrange them very compactly in it, higher in the centre than at the sides, and strew amongst them from three to four ounces of pounded sugar, or more should they be very acid: the grated rind and the strained juice of half a lemon will much improve their flavour. Lay on the cover rolled thin, and ice it or not at pleasure. Send the tart to a moderate oven for about half an hour. This may be converted into the old-fashioned creamed apple tart, by cutting out the cover while it is still quite hot, leaving only about an inch-wide border of paste around the edge, and pouring over the apples when they have become cold, from half to three-quarters of a ping of rich boiled custard. The cover divided into triangular sippets was formerly stuck around the inside of the tart, but ornamental leaves of pale puff-paste have a better effect. Well-drained whipped cream may be substituted for the custard, and be piled high, and lightly over the fruit.

Apple Strudel

—from Florence Kreisler Greenbaum,

The International Jewish Cook Book, 1919

Into a large mixing bowl place one and one-half cups of flour and one-quarter teaspoon of salt. Beat one egg lightly and add it to one-third cup of warm water and combine the two mixtures. Mix the dough quickly with a knife; then knead it, place on board, stretching it up and down to make it elastic, until it leaves the board clean. Now toss it on a well-floured board, cover with a hot bowl and keep in a warm place. While preparing the filling lay the dough in the centre of a well-floured tablecloth on the table; roll out a little, brush well with some melted butter, and with hands under dough, palms down, pull and stretch the dough gently, until it is as large as the table and thin as paper, and do not tear the dough. Spread one quart of sour apples, peeled and cut fine, one-quarter pound of almonds blanched and chopped, one-half cup of raisins and currants, one cup of sugar and one teaspoon of cinnamon, evenly over three-quarters of the dough, and drop over them a few tablespoons of melted butter. Trim edges. Roll the dough over apples on one side, then hold cloth high with both hands and the strudel will roll itself over and over into one big roll, trim edges again. Then twist the roll to fit the greased pan. Bake in a hot oven until brown and crisp and brush with melted butter. If juicy small fruits or berries are used, sprinkle bread crumbs over the stretched dough to absorb the juices. Serve slightly warm.

NOTE: Is apple strudel a pie? Why not? It is encased in pastry, although not in a shaped dish, so it has the same qualifications as a pasty, does it not?

Beef-Steak and Kidney Pie

—from E. Carter, *The Frugal Cook*, 1851

Beat your steaks well, in order to make them eat tender, add one-third the weight of kidneys, cut small in order to extract all the gravy, and season with pepper and salt; line the sides and edge of the dish with paste, cover the whole with your crust, and ornament it as directed for meat pies.

NOTE: Meat and kidney have been used in pies for centuries, but the phrase 'steak and kidney pie' was not in common use until the late nineteenth century. Mrs Beeton (1861) give a recipe for steak and kidney pudding, but the only additions suggested for her basic beef-steak pie are oysters, mushrooms or minced onions. Oysters had been a traditional ingredient in 'beef' pies for centuries, and in Mrs Beeton's day were cheap food, affordable for the poor. It is probable that kidney became the common substitute when oysters started to become expensive as the Victorian era wore on.

Lamb and Currant Pie

—from Cassell's *Dictionary of Cookery*, c. 1875

Cut about two pounds of the breast of lamb into small, neat pieces. Put them in a pie-dish, and sprinkle over them a desert-spoonful of salt, a teas-spoonful of pepper, a tablespoon of finely-minced parsley, a quarter of a nutmeg, grated, and three table-spoonfuls of picked currants. Beat two eggs thoroughly, mix with them a wine-glassful of sherry, and pour them over the meat. Line the edges of the dish with a good crust, cover with the same, and bake in a moderate oven. A little white wine and sugar should be sent to the table with this pie. Time, an hour and a half to bake. Probable cost, 2s. 8d., exclusive of the wine. Sufficient for four or five persons.

NOTE: This pie with its mix of dried fruit and lamb, and the addition of sugar and wine at the table, harks back to medieval times.

Almond Tartlets
—from Theodore Francis Garrett,
The Encyclopaedia of Practical Cookery, c. 1891

Line a dozen tartle-moulds with paste, but the paste on the rims of the moulds, then mask the bottom with a thin layer of marmalade. Pound 6 oz. of blanched Almonds, dried in the oven, mixing up by degrees the same amount of fine sugar, a little orange or lemon zest, and the yolks of six eggs. Remove this from the mortar, put it into a kitchen basin, and work up with it eight whipped whites of eggs. Fill the tartlets, sprinkle them over with fine sugar, and bake in a slack oven for twenty-five to thirty minutes.

Squab pot pie, à l'Anglaise
—from Victor Hirtzler, *The Hotel St. Francis Cookbook*, c. 1919

Roast the squabs and cut in two. Fry a thin slice of

fillet of beef on both sides, over a quick fire, in melted butter. Put both in a pie dish with a chopped shallot that was merely heated with the fillet, six heads of canned or fresh mushrooms, one-half of a hard-boiled egg, a little chopped parsley, and some flour gravy made from the roasted squab juice, and well seasoned with a little Worcestershire sauce. Cover with pie dough and bake for twenty minutes. This is for an individual pie; make in the same proportions for a large pie.

Prune and Raisin Pie

—from Florence Kreisler Greenbaum,
The International Jewish Cook Book, 1919

Use one-half pound of prunes, cooked until soft enough to remove the stones. Mash with a fork and add the juice in which they have been cooked; one-half cup of raisins, cooked in a little water for a few minutes until soft; add to the prune mixture with one-half cup of sugar; a little ground clove or lemon juice improves the flavor. Bake with two crusts.

Lemon Cream Pie

—from Mrs C. F. Level and Miss Olga Hartley,
The Gentle Art of Cookery, 1925

One cup of sugar, one cup of water, one raw potato

grated, juice and grated rind of one lemon. Mix all the ingredients together and bake in pastry top and bottom.

Egg and Bacon Pie

—from Vicomte de Mauduit, *The Viscount in the Kitchen*, 1933

This is a Yorkshire dish par excellence.

Line a pie dish with flaky pastry, cover the bottom with two layers of lean bacon, and break carefully three eggs over them. Salt and pepper, then put the dish uncovered in a hot oven. When the eggs have set, put over them two more layers of bacon, and over this break three more eggs. Pepper, cover with pastry, egg it, and bake in the hot oven till the pastry is cooked.

Vegetable Pie

—from Baron Brisse, *366 Menus and 1200 Recipes,* originally published in France in 1868; this recipe is from the 1896 English translation.

Cook some green peas, young broad beans, small carrots, and tender French beans, separately, in cream sauce; place in a baked pie-case, divided into compartments with thin pieces of paste, and serve. In winter preserved vegetables may be used for the pie. We have to thank the celebrated Grimod de la Rèyniere for inventing this dish, which not only pleases the palate but the eye, and is a great addition to our fast-day dinners.

Select Bibliography

Albala, Ken, *Food in Early Modern Europe* (Westport, CT, 2003)

David, Elizabeth, *English Bread and Yeast Cookery* (London, 1979)

Davidson, Alan, *The Oxford Companion to Food* (Oxford and New York, 1999)

Drummond, J. J., and Anne Wilbraham, *The Englishman's Food* (London, 1958)

Fernandez-Armesto, Felipe, *Food: A History* (London, 2001)

Flandrin, Jean-Louis and Massimo Montanari, *Food: A Culinary History*, trans. Albert Sonnenfeld (New York, 1999)

Hartley, Dorothy, *Food in England* (London, 1954)

Heiatt, Constance B., *An Ordinance of Pottage* (London, 1988)

Hess, L. John and Karen, *The Taste of America* (New York, 1977)

Kiple, Kenneth F. and Kremhild Coneè Ornelas, *The Cambridge World History of Food*, vols I and II (Cambridge and New York, 2001)

Larousse Gastronomique (New York, 2001)

Root, Waverley, *Food* (New York, 1980)

Shephard, Sue, *Pickled, Potted and Canned* (London, 2000)

Stewart, Katie, *Cooking and Eating* (London, 1975)

Toussaint-Samat, Maguelonne, *History of Food*, trans. Anthea Bell (Oxford, 2000)

Trager, James, *The Food Chronology* (London, 1995)

Websites and Associations

Culinary & Dietetic Texts of Europe from the
Middle Ages to 1800:
www.uni-giessen.de/gloning/kobu.htm

Gode Cookery
www.godecookery.com/godeboke/godeboke.htm

Historic Food
www.historicfood.com/portal.htm

The Food Timeline
www.foodtimeline.org/

The Historic American Cookbook Project:
http://digital.lib.msu.edu/projects/cookbooks/

What's Cooking America: History and Legends of
Favorite Foods:
http://whatscookingamerica.net/History/HistoryIndex.htm

The American Pie Council:
www.piecouncil.org/national.htm

The Melton Mowbray Pork Pie Association:
www.mmppa.co.uk/about.html

Official Great Aussie Meat Pie Competition
www.greataussiepiecomp.com.au

图书在版编目（CIP）数据

派：汉英对照 /（澳）珍妮特·克拉克森著；李天蛟译.
——北京：北京联合出版公司，2024.3
（食物小传）
ISBN 978-7-5596-7321-3

Ⅰ．①派… Ⅱ．①珍… ②李… Ⅲ．①面食-文化史-世界-普及读物-汉、英 Ⅳ．① TS972.132-49

中国国家版本馆 CIP 数据核字（2023）第 244292 号

派

作　　者：〔澳大利亚〕珍妮特·克拉克森
译　　者：李天蛟
出 品 人：赵红仕
责任编辑：孙志文
产品经理：夏家惠
装帧设计：鹏飞艺术
封面插画：〔印度尼西亚〕亚尼·哈姆迪

北京联合出版公司出版
（北京市西城区德外大街 83 号楼 9 层　　100088）
北京天恒嘉业印刷有限公司印刷　　新华书店经销
字数 113 千字　889 毫米 ×1194 毫米　1/32　8.25 印张
2024 年 3 月第 1 版　　2024 年 3 月第 1 次印刷
ISBN 978-7-5596-7321-3
定价：59.80 元

版权所有，侵权必究
未经书面许可，不得以任何方式转载、复制、翻印本书部分或全部内容。
本书若有质量问题，请与本公司图书销售中心联系调换。电话：010-85376701

Pie: A Global History by Janet Clarkson was first published by
Reaktion Books, London, UK, 2009, in the Edible series.
Copyright © Janet Clarkson in 2009
Rights arranged through CA-Link International
Simplified Chinese translation copyright © 2024
by Phoenix-Power Cultural Development Co., Ltd.
ALL RIGHTS RESERVED

版权所有 侵权必究
北京市版权局著作权合同登记 图字：01-2023-2070 号

Acknowledgements

This book has been a great deal of fun in the making. I am delighted to have been given the opportunity to delve into and write about a topic so dear to my heart and heritage, and am honoured to be included amongst the illustrious company of the other authors in the Edible series.

My gratitude goes first to Andrew F. Smith, culinary historian and author, for recommending me for the project, to Michael Leaman of Reaktion Books for accepting me on Andrew's recommendation, and to Martha Jay for her patience with me during the post-writing production phase.

Special thanks are due to my son-in-law Patrick Bryden for his invaluable help with sourcing and preparing many of the images for this book, and to my husband Brian and our special friend Trevor Newman for managing to hold off eating the Melton Mowbray and *Pézenas* pies long enough to photograph them. I am also appreciative of the great body of loyal readers of my blog *The Old Foodie*, who have reinforced my belief that food history can be fascinating for the 'general reader' as well as the historian.

Finally, I am indescribably grateful for the unfailing support and unflagging enthusiasm of my friends and family, especially that of my husband Brian, my children (and their partners) Matthew (and Vicki) and Sarah (and Patrick), and my little sister, Val. Thank you from the bottom of my heart.

Photo Acknowledgements

The author and publishers wish to express their thanks to the below sources of illustrative material and/or permission to reproduce it. Locations of some artworks are also given below.

Accademia di Belle Arti di Brera, Milan: p. 010; photo Brian Clarkson: p. 013; photo Sandra Cunningham/shutterstock images: p. 109 (foot); photo © Janis Dreosti/ 2008 iStock International Inc.: p. 090; photo Everett Collection/Rex Features: p. 097; photo © Jack Jelly/ 2008 iStock International Inc.: p. 050; photos Library of Congress, Washington, DC: pp. 019, 022, 058(foot), 067, 083, 109(top right); photo © Liza McCorkle/ 2008 iStock International Inc.: p. 053; Maidstone Museum & Art Gallery, Kent: p. 027; photo © Monkey Business Images/shutterstock images: p. 109 (top left); Musée d'Orsay, Paris: p. 020(foot); Musée du Louvre, Paris: p. 044; photo Trevor Newman: p. 047; photo © Jeanell Norvell/ 2008 iStock International Inc.: p. 2(contents page) ; photo Roger-Viollet/ Rex Features: p. 007; photo courtesy of Evan Schoo: p. 085; Tate, London: p. 087; © D. C. Thompson & Co., Ltd.: p. 006; from *The Times* (June 1945): p. 056(right); photo courtesy http://www. cornishlight.co.uk/: p. 069; courtesy www.weebls- stuff.com: p. 104.